*f*P

A
GRAND
AND
BOLD THING

An Extraordinary New Map of the Universe
Ushering In a New Era of Discovery

ANN FINKBEINER

Free Press

NEW YORK LONDON TORONTO SYDNEY

Free Press
A Division of Simon & Schuster, Inc.
1230 Avenue of the Americas
New York, NY 10020

First Free Press hardcover edition August 2010

FREE PRESS and colophon are trademarks of Simon & Schuster, Inc.

For information about special discounts for bulk purchases,
please contact Simon & Schuster Special Sales at 1-866-506-1949 or
business@simonandschuster.com.

The Simon & Schuster Speakers Bureau can bring authors to your live event.
For more information or to book an event, contact the Simon & Schuster
Speakers Bureau at 1-866-248-3049 or visit our website at
www.simonspeakers.com.

Manufactured in the United States of America

1   3   5   7   9   10   8   6   4   2

Library of Congress Cataloging-in-Publication Data

Finkbeiner, Ann K.
A grand and bold thing: an extraordinary new map of the universe ushering in a
new era of discovery / Ann Finkbeiner.—1st Free Press hardcover ed.
        p.   cm.
1. Three-dimensional imaging in astronomy. 2. Astronomical spectroscopy.
3. Astronomy—Charts, diagrams, etc.—Data processing. 4. Astronomical
instruments—Technological innovations. 5. Galaxies—Observations.
6. Quasars—Observations. 7. Gunn, J. E. (James Edward), 1938– I. Title.
        QB465.F56 2010
        520.22'3—dc22
        2010008533

ISBN 978-1-4165-5217-8

For Papa and T.C.,
with all my love

Contents

A

GRAND

AND

BOLD THING

The Instrument Fairy

"And what do you want to do with the new telescopes?" I asked Wallace Sargent, who's a Caltech astronomer. "That's a boring question," he said, "and the answer is so boring I won't answer it. It's like you're a Victorian explorer looking for the source of the Nile and when you run across the Pyramids, if you had any sense at all, you'd investigate them. Pardon the expression, but I point the fucking telescope at the sky and see what's out there."

—interview with Wallace Sargent, 1991

THE HUBBLE SPACE TELESCOPE'S first image looked like all hell. A month later, in late June 1990, the telescope's political shepherd, John Bahcall, found out what had gone wrong. He called some interested local astronomers—Jill Knapp, Don Schneider, and particularly Jim Gunn—and since astronomy in Princeton, New Jersey, usually involves food, he invited them to supper at a Route 206 strip-mall Chinese diner. He told them that NASA was about to announce that the telescope's perfect mirror had been ground to the perfectly wrong shape. Jim Gunn, who had designed and overseen the construction of the telescope's principal camera, had also seen the first image and had thought the problem might be fixed. But no, now Bahcall was telling him no, it was the mirror's shape, the telescope couldn't focus, the problem was irrevocable.

Jim and Bahcall did most of the talking. Why hadn't the astronomers overseeing the NASA contractors caught the mistake? It was a dumb mistake, a beginner's textbook mistake. Nor did they think that astronauts could change out the mirror on a telescope in orbit nearly 400 miles up. Could they themselves come up with any brilliant ideas? Not at the moment. Certainly any kind of repair would take years, money, and a lot more effort, and they both had already dumped their lives and careers into it. Don Schneider, who had been Jim's graduate student and was now working with Bahcall, sat there feeling overwhelmed. Jill Knapp, who was married to Jim, mostly listened. Dinners with Bahcall always started early, so they all went home early too.

Jim Gunn had gotten involved with the planning for the Hubble Space Telescope fifteen years before, when he was thirty-eight years old and a full professor at Caltech. His own reason was that he and most other cosmologists—those astronomers who study the origin and evolution of the universe—could use the telescope to find out how the universe would end. Immediately after the Big Bang, the explosion that began it all, the new universe had begun expanding, and now, thirteen-plus billion years later, it was still expanding. No one knew how fast it was expanding because that depended on how much matter was in it, and they didn't know the amount of matter because at least 90 percent of it was invisible; they called it dark matter. Enough of any kind of matter, and its mutual gravitation would make the universe eventually slow, stop, reverse direction, and collapse in on itself. Not enough matter, and the universe would just get more rarefied and expand forever. The amount of matter in the universe and the rate of expansion so preoccupied cosmologists they went around saying that all cosmology was just the search for two numbers.

At the same time, Jim happened to be fixed on a newly invented light detector called a charge-coupled device, or CCD, which takes light and turns it into digital information. CCDs are unusually sensitive. In fact, they were thirty times more sensitive than the light detector that Jim knew was planned for the Hubble Space Telescope's main camera. So in October of 1976, Jim had walked into the office of his friend and colleague Jim Westphal in Caltech's South Mudd building and said, "Jim, we have to build the camera for the space telescope." A CCD camera on a space telescope would be sensitive enough to find cosmologists' two numbers.

But building a camera for the Hubble Space Telescope meant working through NASA, and for Westphal and Gunn, the prospect was unappealing. The NASA style is corporate, takes decades, costs hundreds of millions of dollars, and operates on systems engineering principles, one of which is to do things only well enough to get them done. This is the polar opposite of astronomers' style, which is iconoclastic, takes months to years, is as cheap as possible, and operates on the principle of Westphal and Gunn in the basement with soldering irons doing things right, period.

So back in South Mudd, Westphal told Gunn he wanted to have nothing to do with NASA. But Gunn, gripped by a vision of sensitive electronics, went to the blackboard and wrote the equations that showed how CCDs in space meant seeing better by a factor of one hundred. Westphal wasn't convinced: "Jim, no way," he said. "We'll build some nifty thing and the damned thing will have a bad solder joint in it someplace and we won't be able to reach up there and fix it, and it'll drive both of us into the booby house."

Gunn became eloquent: "If we don't do it, we're going to be out of business in astronomy, in serious astronomy, the forefront of astronomy, on the ground, in five years." So for the next ten years, Gunn, Westphal, a team of astronomers, and a large team of NASA contractors built what came to be called the Wide Field and Planetary Camera, or WFPC, pronounced "wiffpick." Astronomers said that WFPC would make the whole space telescope worthwhile.

But sure enough, Westphal had been right about NASA. A month after the space telescope's launch, WFPC took the first picture with the bad mirror, and what should have been a sharp little star resolved to a brilliant point was instead a sprawling, streaky, blurry mess. NASA's investigation into what went wrong blamed everyone: NASA wasn't overseeing the aerospace contractor, the contractor wasn't overseeing its own engineers, and the astronomical community wasn't overseeing either one.

The astronomers felt they'd been screwed over. Gunn didn't think they had. He thought the whole point of the telescope was astronomy, and astronomers hadn't bothered to get their hands dirty making it. They were elitists—"gentleman astronomers in their coats and ties," he called them—who rarely knew much about their own instruments and left the mundane and picky technical work to the mere technicians. He and Westphal had braved NASA's culture and worked closely with WFPC's contractor, and he thought the astronomers should have been hovering over NASA, harping about the details. He really felt that some people believed instruments were brought by the instrument fairy.

In July 1990, he sat down to write a letter to his community.

"My view and point are very simple," he began. "We were not 'screwed over'—we have been exquisitely vulnerable to precisely this kind of thing happening for years." He explained the vulnerability: "We are a discipline of technical incompetents, happy to let our or NASA's engineers build our tools to their desires, by and large, not ours." He began using capital letters and took a shot at NASA: "It was an ASTRONOMICAL failure; it was an ASTRONOMICAL satellite, and it does not matter a whit that it was probably some fool at PE [PerkinElmer, the NASA contractor for the mirror] that caused it and some entirely expected failure of NASA's criminally infantile QA program that failed to catch it." Then he took dead aim at the astronomers: "Not that there would have been enough technically competent people in the field to have taken the two G$ and done anything with it." He concluded: "You may certainly circulate this letter if you wish."

The letter spread electronically—astronomers had early on embraced e-mail—and it was printed out and taped to office doors, forwarded and posted to electronic bulletin boards, and talked about in the halls of astronomy departments everywhere. Afterward Jim decided to cut bait; he gave up his privileges to use the space telescope. "I was just terribly, terribly disheartened," he said. "I had immersed my life in it, I neglected important things. I needed to start over." He didn't think it had a chance in hell of being fixed anyway.

NASA, of course, pulled off an ingenious fix, installed by astronauts dangling improbably over the telescope up in space. The Hubble Space Telescope went on to find cosmology's two numbers and took gorgeously sharp pictures of everything else: of tiny, distant, golden galaxies and nearby electric blue whirlpools, of radiant towers of star birth and the opalescent mists of their deaths. An upgraded version of Jim's and Westphal's WFPC took some of the best of them.

_____

Jim regretted having cut bait, but only a little. He had just come up with another idea for another camera, even better than WFPC, with a wider field of view and bigger, more sensitive CCDs. He'd put the camera on an unassuming telescope on the ground, turn it on, and let the night sky roll over it. It would make the first-ever three-dimensional digital sky survey.

The survey would have a million galaxies in it. With a million galaxies you could make a map showing where galaxies lived, and whether they lived around other galaxies, and whether little chains of galaxies were actually parts of a larger network, and whether that network was actually a network of networks, and how big it all got. With a million galaxies you could make a true census: how galaxies differed, how baby galaxies changed as they grew, what kinds of galaxies they grew into. With a million galaxies you could watch the universe growing up.

Jim thought a million-galaxy survey was intellectually huge and personally irresistible, and he was pretty sure he could do it without running into the large teams, cost overruns, time delays, and management problems that the Hubble had had. It could be done astronomer style, simply and cheaply and right. It would create the most complete map of the universe ever, and it would be accessible at all hours to all comers for all time.

Chapter 1

Stakes Worth Playing For

I earned my spurs by doing other stuff, but instruments are what I enjoy.

—Jim Gunn, Princeton University

JAMES EDWARD GUNN is thin and pale and below medium height; his hands are outsized and look remarkably competent. He has mild, dark eyes and a look of remote sweetness. Sometimes he goes into trances and you can't tell whether he's concentrating on something he's looking at, or thinking about something else, or just in a state of sleep and energy deprivation. For some reason, he almost always wears Hawaiian shirts.

With nonastronomers, he is patient with ignorance and polite to the point of deference. He's patient and polite with astronomers too, though now and then he gets wrought up at what seems to him to be willful stupidity or inattention to exactness. Once wrought up, he continues to talk politely but with increasing intensity, and some words are in aural italics or, if he's writing, in capital letters. Sometimes he ends up saying something he shouldn't. When he's not traveling, he spends most of his time in the basement of Princeton University's Peyton Hall, in a crowded but highly organized electronics workshop. In the middle of the workshop is a circular six-foot vacuum chamber that has been obsolete for years but is too big to be moved out through the door; Jim put his desk behind it, where, from the door, he's invisible.

Jim is famous. He has a named chair in one of the world's best astronomy departments and has won nearly every prize available to an astronomer, including a MacArthur and the National Medal of Science. Astronomers admire him for his ability to build astronomical instruments, use them to make observations, and explain the observations with theory. Theorists claim Jim as a theorist and observers claim him as an observer. He's that good at everything.

Jim was born in 1938 in Livingston, a small town in the oil fields of east Texas. His father was an exploration geologist who moved to new oil fields every six months or so; Jim grew up in Mississippi, Alabama, Georgia, Arkansas, Louisiana, Oklahoma, Texas, and Florida. With all that moving and because he was an only child, he was pretty much a loner. His closest friend was his father, who had set up a machine shop in a trailer that moved with them—during the war, when parts were hard to get, his father had to make them himself. From his father, Jim learned the construction of intricate things.

Jim was five years old when he read *The Stars for Sam*, which, while written for children, was not written for five-year-olds. The book said that all the stars in the sky were in our galaxy, the Milky Way, and that the Milky Way was only one of many galaxies, all filled with their own stars and scattered throughout space. Each one of those galaxies could be called an island universe. All together the universe held 30 million of them, and they could be classified by shape—spiral, elliptical, etc.—and cataloged as in a museum. When Jim was six, he and his father sent away for lenses and built a telescope. The next year, Jim found his father's old undergraduate astronomy textbooks and read them, fascinated by the idea that the universe had a beginning.

When he was twelve, his father died suddenly of heart disease and Jim felt part of his life had emptied out permanently; he didn't like to think about it. So he focused his energies on power tools. He designed and built twenty or thirty model airplanes, a couple of telescopes, furniture, a hi-fi, some rockets, and played around with high explosives.

High school was mostly in Beeville, Texas, where he was astonishingly good at science and math, and where he read *Frontiers of Astronomy* by the British astronomer Fred Hoyle. Hoyle hadn't liked the theory that the universe was created in an explosion and had named it, snippily, the Big Bang theory: "an explosive creation of the Universe is not subject to analysis," he said with some justification. Hoyle's own theory, laid out in his book, was that the universe lives, and has always lived, in a balanced and steady state—an excellent idea, Jim thought. Hoyle went on to say that everything in the universe was related: "From the vast expanding system of galaxies down to the humblest planet, and to the creatures that may live on it, there seems to be a strongly forged chain of cause and effect." All astronomical evidence could fit into one framework, Hoyle wrote, and that framework was the laws of physics. Since those laws were comprehensible, the universe was comprehensible too.

For a high school student, this was powerful stuff. The laws of physics are a set of rules you can believe, you can build on, and they hold for every place in the entire universe. And all the pieces of the universe—the earth, the planets, the sun, the lives of all kinds of stars, the galaxies, the universe's expansion—were intimately related, and so they should click together into one coherent picture. But Hoyle warned that lovely as this picture is, we shouldn't believe it until we have tested it in every way. The picture, the theory, must be exposed to observational attack from every direction and still endure. Only then, wrote Hoyle, would we be in a position to obtain a complete understanding of the universe as a single, interlocking thing. "The stakes are high," he wrote, "and win or lose, are worth playing for." Jim decided to be an astronomer.

For college, Jim stayed in Texas and went to Rice University in Houston, which was the best university in that part of the world. He'd been accepted at other schools, but Rice charged no tuition for local students and Jim's family—his mother had remarried— was not well-off. To cover living expenses, he won scholarships and achievement prizes. Rice didn't offer a major in astronomy, but Jim decided that was fine: instead, he'd major in astronomy's foundations, math and physics. Besides, he thought, while astronomy is easy to learn, physics is hard and you can't pick it up on your own. As it turned out, he found he was flat-out good at it. But he hadn't stopped reading astronomy books, and the summer before he went to Rice, he'd built a little 8-inch reflector telescope for which he'd ground and polished the mirror; while he was at Rice, he added a motor to drive the telescope, then a camera to take pictures of what the telescope saw. He finished it during his senior year and wrote it all up for the observers' column in *Sky & Telescope*.

In 1961, the year he graduated summa cum laude, *Time* ran a feature called "Top of the Heap" about some "extraordinary" college graduates, and Jim was one of them. Jim studied ten hours a day, *Time* said, and his only extracurricular activity was astronomy club. His physics professor was quoted, saying, "I've never been able really to determine the limits of his ability. I've never been able to ask him an exam question that he can't give a perfect answer to." Because scientists have to get doctoral degrees—the *philosophiæ doctor*, the PhD—and because Jim was now fascinated by physics, he applied to and was accepted at the astronomy department at the California Institute of Technology to study Einstein's general relativity, the physics that was the basis of all studies of the universe.

Caltech looks like a garden—jewel-green grass in brown, dry Southern California, flowers and flowering trees, winding paths, little fountains and pools and waterfalls—set among pale stuccoed buildings with red tile roofs, connected with arcades. It's light and graceful and silent. At Caltech, Jim felt that he'd landed in the midst of an intense community of scholars who knew the answers to his questions, with whom he had no trouble being heard, who treated him as one of their kind. He felt almost light-headed with joy.

Happily, Caltech was also rich. It owned several telescopes, which were situated about 100 miles south of campus on Palomar Mountain, and one of them, the 200-inch Hale Telescope, was the biggest in the world. With a bigger telescope, you can see things that are fainter and farther away; the farther away, the farther back in time. The early universe had for years been the province of theorists only; until recently nobody had actually seen it. When Jim got to Caltech, Caltech's astronomical observers had just found the earliest and most distant things anyone had ever seen.

These distant things happened to be sources of intense radio waves—Caltech also owned a radio telescope. And because astronomers have a human prejudice toward the optical wavelengths in which they see, Caltech observers had double-checked the radio sources with optical telescopes. The radio sources were outright odd. They looked like pinpricks the way stars do, but the other information about them had been unintelligible. A Caltech observer named Maarten Schmidt had just figured out that the radio sources made sense only if these starlike things were at highly unstarlike distances.

An object's distance can be read from its velocity—how fast it's moving with the universal expansion—and its velocity can be read from its spectrum, its light spread by a prism into a rainbow. Superimposed on the rainbow are specific features—they can look like the lines of a barcode and are caused by the behavior of the galaxy's atoms—at specific colors, that is, specific wavelengths. When the galaxy is moving away, its light takes longer and longer to get to us, and these features in the spectrum shift down toward longer, redder wavelengths; astronomers call this redshift. The larger a galaxy's redshift, the faster it's moving away from us and the more distant it is. Schmidt had already found a regular galaxy with a redshift of 0.46, meaning that it was about 5 billion light-years away.

Then Schmidt noticed that one of these radio sources, called 3C 273—with a redshift of 0.16, only about 2 billion light-years away—was enormously brighter than the redshift 0.46 galaxy. So Schmidt and his colleagues looked at the other odd radio sources and found more single starlike things, each forty times brighter than the biggest galaxies full of 10 billion stars. They named the starlike things quasi-stellar radio sources; later they shortened the name to quasars. In the next few years, they found a handful of quasars, each one farther away than the one before. No one had any idea what quasars were or what made them shine so brightly.

In 1965, when Jim was a fourth-year graduate student and hardly aware of quasars, he went to a talk Schmidt gave. Schmidt began with his first quasar and walked the audience through the rest one by one: 3C 254, redshift 0.73; 3C 245, redshift 1.02; 3C 287, redshift 1.05; 3C 9, redshift 2.01—10 billion light-years away, two-thirds of the way back to the beginning of the universe. These redshifts, these distances, these ages, were unheard of. Schmidt himself, who wasn't yet forty years old, had only dreamed of getting such redshifts by the end of his career. Jim thought it was the most exciting single astronomy talk he'd heard.

Jim had gone to the talk with a fellow graduate student, Bruce Peterson, and sitting there listening, they noticed an anomaly in the spectrum of the quasar 3C 9. 3C 9 was in the early universe; the universe between 3C 9 and Caltech was full of hydrogen atoms, and hydrogen atoms absorb light at certain ultraviolet wavelengths. At those wavelengths, then, 3C 9's spectrum should have been flat and dark, but it wasn't; it was full of ultraviolet light. That light would be there, Jim and Bruce thought, only if something had changed all those hydrogen atoms—if their electrons had been ripped off, or ionized, by something violently energetic or hot. If so, that violence must have been coming from something like giant stars lighting up or quasars, whatever they were, doing whatever they did.

The dark place in the spectrum that should have been there became known as the Gunn-Peterson trough or the Gunn-Peterson effect. Finding an object whose spectrum showed the trough would mean finding an object from the time when the hydrogen atoms were still intact, before the universe created the first energetic, shining objects, when the universe was an infant. But the trough wasn't there in 3C 9 at redshift 2.01, and no one knew how much farther back you'd have to go to see it—certainly farther than their instruments could now see. Gunn and Peterson published their idea in 1965 and because it was so clever, they became mildly famous for it. Jim was happy to have a trough named after him and wished someone would find it, but in the meantime he went on to other things until the day when instruments improved.

After Jim graduated from Caltech in 1966, he spent the next few years in the army, building instruments at NASA's Jet Propulsion Lab—Caltech's astronomers had connections that could keep their star graduate students around telescopes—and then took a position as an assistant professor at Princeton. For a while, he worked with Princeton astronomer Jeremiah Ostriker on a theory about an unusual kind of regularly flashing star just discovered, called a pulsar. But Princeton didn't have a telescope, and Jim felt he needed one, so in 1970 he returned to Caltech.

For the next ten years he stayed at Caltech, working meticulously and painstakingly on one research project after another, heading off on one subject, then digressing onto another—usually without giving up the first—and then digressing yet again. His style of research was uncommon. Most astronomers stake out claims on certain kinds of objects or certain parts of the universe: for example, Maarten Schmidt had spent much of his career studying quasars. But Jim worked on different populations of stars; the evolution of stars; the Milky Way; gravitational lenses; the particles of dark matter; binary stars; local galaxies; rare and peculiar stars; globular clusters; supernovae; quasars; and clusters of galaxies. Some of these subjects are related—classifying stars into populations can mean tracking the early, middle, and late stages of their evolution—but most are not. And all the while he was inventing and building cameras and spectrographs. He obviously had an enormous intellectual range and apparently the attention span of a housefly.

In fact, he seemed to have taken Hoyle's interconnected universe to heart, studying each of the universe's pieces to see if somehow they'd click together into one whole history. Toward that end, he began two long-term surveys of the universe's opposite ends.

Maybe if you could connect the near and far, the present and past ends of the universe, you could see how it changed with time and you could figure out its history. He started with the far end, the earliest things seen, the quasars.

Maarten Schmidt had been racking up higher and higher numbers of quasars, and the highest numbers seemed to be at the greatest distances. Other quasar hunters had joined him and found the same. Then, somewhere around redshift 3.0 or 4.0, the numbers dropped dramatically; quasar hunters called the drop the quasar cut-off. It seemed to imply that somewhere out beyond redshift 3.0, at even higher redshifts even earlier in the universe, quasars first appeared.

Before believing the cut-off, though, you'd want to expose it to Hoyle's observational attack and systematically count quasars at increasing redshifts. The quasar hunters had collected a few hundred quasars, but believability required better statistics, and those wouldn't be achieved by collecting quasars one by one. One night in 1977, Maarten Schmidt and Jim were having dinner at Palomar, and Schmidt asked Jim whether the new CCD camera Jim had just built, a prototype for WFPC, could be rigged to do a wholesale spectroscopic survey of quasars.

Within a year of dinner with Schmidt, Jim had built a CCD camera that was sensitive to the faintest light. He called it the Prime Focus Universal Extragalactic Instrument, or PFUEI (pronounced the way it looks, "phooey"). PFUEI was mounted at prime focus, near the top of the telescope, and to use it Jim sat in a little cage next to it, 100 feet off the ground. Sitting up there was cold but glorious, and just that 100 feet closer to the stars. He'd put PFUEI on the Palomar 200-inch, point it at the sky, lock it in place, and let the earth's rotation move the sky past the CCDs—

a maneuver called drift scanning. In the early 1970s, getting the spectrum and therefore the redshift of one quasar the traditional way could take most of the night. In the same amount of time, PFUEI could find the redshifts of twenty to thirty quasars.

Jim, Schmidt, and Don Schneider, who had been Jim's graduate student and was then Schmidt's postdoc, didn't actually begin the survey with PFUEI until the early 1980s. They found no quasars with redshifts over 3.0, and they had expected to find many—an interesting contradiction, but as scientists like to say, absence of evidence is not evidence of absence, so they had no proof of a cut-off. In the mid 1980s they expanded the survey, this time with a second camera Jim built with four CCDs arranged in a square, which Jim named Four-Shooter. It could see much more sky and was much more sensitive. Over the next few years, the Four-Shooter survey found one hundred high-redshift quasars, including a record setter at redshift 4.04, then one at 4.73, then another at 4.9. Schmidt, Schneider, and Gunn considered they had a reasonable suggestion of a cut-off and published: beginning nearby, the numbers of quasars rose until, around redshift 3.0, only 2 billion years after the Big Bang, they peaked, a hundred times more quasars than now. Even farther back than that, their numbers seemed to fall rapidly, or at least get harder to see and count, but in any case, quasars turned on no later than redshift 4.9. They still hadn't seen the Gunn-Peterson trough; if it existed, it was farther back than the farthest quasar.

During much of the time Jim was surveying quasars in the distant universe, he was also working with another Caltech colleague, John Beverly Oke, on the other survey, the one of the nearby universe. They were looking specifically for local evidence of cosmology's two numbers, the universe's density and its rate of expansion.

The exact rate at which the universe expands had been cosmologists' problem child since 1929, when Edwin Hubble measured the distances of some nearby galaxies—he called them nebulae—and found that the farther the galaxies, the faster they were moving away from us, as though they were painted on the skin of an inflating balloon, as though they were riding expanding space. The expansion rate became known as the Hubble constant; Hubble thought it was 500 kilometers per second every 3.26 million light-years, that is, every megaparsec.

Ever since, observational astronomers had been measuring and remeasuring the expansion and disagreeing wildly with one another's numbers. One camp said the constant was 50, the other said it was 100, and Jim didn't believe either one. In any case, he was more interested in the universe's density, in whether the amount of its gravitating matter could slow the expansion enough that the universe would collapse in on itself, or whether it would just coast on, expanding forever.

Back in 1974, Jim and three other young cosmologists—Richard Gott of Princeton, David Schramm of the University of Chicago, and Beatrice Tinsley of the University of Texas—had put together all the physics, observations, interpretations, and arguments about the known state of the universe, assessed them all for believability, and in a scholarly paper called "An Unbound Universe?" announced that the universe, though slowing, would never stop expanding. Their paper became known for being succinct, exhaustive, and brave, and it was eventually referred to just by names of the authors: Gott, Gunn, Schramm, and Tinsley. The question was, were they right?

At the time, astronomers were mostly measuring expansion, and to do that, they balanced velocity—that is, redshift—and distance. Redshifts could by now be measured with fair precision, repeatability, and lack of controversy. Measuring distances, however, was a hornet's nest and the reason the Hubble constant fight was so bitter. Distance can be extrapolated from brightness: the more distant an object is, the dimmer it appears. This measure works only with certain families of objects whose brightness is so standardized that astronomers call them standard candles. Brightness falls off proportional to the distance squared: so if standard candle 1 is a hundredth as bright as standard candle 2, then standard candle 1 is reliably ten times more distant.

For years, astronomers tried one standard candle after another and continued to disagree wildly. Their measuring errors were large, between 10 and 30 percent. The candles that were reliably standard were too nearby to say much about the expansion of the entire universe, and the ones far enough away couldn't be trusted to be standard.

The standard candle Jim wanted to work on, and the one most popular in the early 1970s, was called the brightest cluster galaxy. In clusters of galaxies, the galaxy in the center is usually the biggest and brightest. Astronomers called the big, bright, central galaxy a brightest cluster galaxy, or just a BCG. The ones nearby all had about the same absolute brightness and should make good standard candles. So when Jim and Bev Oke began measuring the universe's expansion, they surveyed first for clusters, and then within the clusters for the BCGs. By the mid-1970s, they'd found enough BCGs to estimate that the universe's expansion was in the same range as Gott, Gunn, Schramm, and Tinsley had predicted.

But Jim didn't trust the results. Beatrice Tinsley had just given a talk at Princeton saying that BCGs were unreliable: while nearby BCGs were one kind of standard candle, the distant ones almost certainly had to be another kind. Galaxies are made of stars. Stars evolve; as they age, they change color and brightness. Obviously galaxies must evolve too, changing color and brightness, and obviously they're useless as standard candles. Jim was at Tinsley's talk with two other cosmologists, Jeremiah Ostriker (with whom Jim had worked on pulsars) and Scott Tremaine, who were now inspired to make the uncertainty even worse.

Ostriker and Tremaine calculated the paths of all the galaxies within a given cluster. Inside clusters, galaxies buzz around like flies, colliding, merging, generally swapping around stars. At the same time, they are orbiting the BCG in the center, gradually slowing, spiraling in until the BCG captures them, one after another. So not only are the stars in the cluster galaxies evolving, the cluster galaxies themselves are merging with one another and with the BCGs, changing the BCGs' brightness. BCGs once and for all lost the case for being standard candles. Jim got disenchanted. He'd thought the problem would be simple, or at least straightforward, and here was Nature one-upping him, saying prissily, "Everything is a great deal more complicated than you think."

But Jim and Bev Oke had this perfectly good cluster survey, and they didn't want to give up on it. If BCGs in the clusters couldn't be used to find the universe's fate, maybe the whole clusters themselves could. Cosmology's two numbers, expansion and density, are linked—high density always means slowed expansion, and vice versa—and by measuring one of them, you get the other for free. So forget expansion and standard candles; try the other number, try density and gravity instead.

Clusters of galaxies are the largest objects in the universe that are held together by gravity, and the amount of time gravity would take to pull them together would depend on the overall density of the universe. In a high-density universe, clusters trying to pull themselves together will have more competition from surrounding matter and will take longer to form. So at a higher redshift—the earlier, more distant part of the universe—clusters should be fewer in number. Jim and Oke and a variety of collaborators decided to continue their cluster survey, counting the number of clusters at greater and greater distances. By the mid-1980s, they'd found enough clusters between redshifts 0.5 and 1.0 that they could again support Gott, Gunn, Schramm, and Tinsley's vote for universe with low density and slow and infinite expansion.

And again, Jim couldn't quite trust the observations, and again, the reason was that no one knew quite how cluster galaxies changed with time. They were counting clusters down to some minimum brightness. But maybe the galaxies in the clusters had been brighter in the past; if so, they should be able to see more of them farther back. And high numbers of clusters in the early universe wouldn't reliably imply the universe's density after all.

The one outcome of the cluster survey that Jim did trust implied the general course of the universe's history. Jim and collaborator Alan Dressler studied several clusters in detail, comparing the younger, distant ones with older, nearby ones. The farthest clusters, at redshift 0.5, seemed to have a high fraction of galaxies that were either bursting with newly forming stars or clearly just poststarburst. The nearest clusters, redshifts of 0.02 or so, seemed full of galaxies that were dead, whose stars had gone from infant blue through adult yellow to old-age red. The universe seemed to be turning from hot, young, and blue to cool, old, and red—an unverified and broad-brush history, but a history nonetheless. The universe had been a much livelier place when it was young.

About now Jim began to feel that he was spinning his wheels. He'd done an extraordinary amount of work on instruments, theory, and observations in an extraordinary number of fields. But he worried that he'd begun operating on inertia. He was in a complex state of mind. The quasar and cluster surveys had been important, and his colleagues were right in continuing them, but in his own mind, the surveys had gone about as far as PFUEI and Four-Shooter could take them, and further observations were hitting diminishing returns. Both surveys covered a few square degrees—a few Moons' worth—and had a few hundred objects chosen because they were bright enough to be seen, and neither the smallish area nor the fewish objects nor the bias toward brightness inspired confidence. Cosmologists should always doubt their observations, Jim thought, and the surveys' uncertainties were no worse than any other astronomer's observations. But even so, too much of what he'd done didn't appear to him to be secure enough to publish, let alone to believe.

In fact, too much of late-twentieth-century cosmology was still not quite believable, more theory than observation. Observers did the best they could, but the observations themselves were time-consuming, finicky, and had measuring errors that could be the size of the measurements themselves. An eminent observer, Vera Rubin of the Carnegie Observatories, said that observers needed to be incredibly devious and take ten objects and infer the rest, but for her to believe anything about the arrangement of galaxies on the large scale, she said, she'd want the redshifts of a million galaxies; the number known at the time was just over ten thousand. An equally eminent theorist, James Peebles of Princeton, said that when his latest theory of how galaxies formed became accepted, he got nervous: "I could think of lots of other ways to make galaxies," he said. Jim told the writer Alan Lightman in 1988, "Cosmological observations are always right at the hairy edge of the possible." As a result, he said, observers overinterpreted their observations and theorists overinterpreted them even more: "The correlation of what is really true about the universe and the set of notions that we think are true about this universe I think is not very high at the moment." Peebles said, "There was a lot to do in the late eighties."

Jim thought they had to start with galaxies. His surveys had made it clear that if you wanted to understand the universe, you had to understand galaxies. Galaxies are the basic units of the universe; all the stars are in galaxies. At the far end of the universe, quasars—astronomers now knew—were things that happened in galaxies, and to understand quasars, you had to understand galaxies. And at our near end of the universe, to understand clusters, you had to understand galaxies. You couldn't find the expansion, density, or fate of the universe and you couldn't map its history at all without understanding galaxies. And so little was known about them. Galaxies seemed to come in a whole zoo: blue, red, faint, bright, spiral, elliptical; galaxies that looked like messes, or that were forming stars actively, or that were just sitting there; massive galaxies, wispy little ones, galaxies living in clusters or scattered loosely around in empty fields.

In particular, Jim thought, cosmologists' affection for the distant universe meant that no one had a reliable census of the galaxies nearby. How could you build a history if you studied only the past and never found out what the present looks like?

In fact, if the universe was a single thing whose parts were all related and which evolved with time, then cosmologists weren't going to see it. Cosmologists had, from technological necessity, broken the universe into different, apparently unrelated problems: expansion and density and fate, quasars and the earliest universe, galaxy clusters and galaxies. Jim's career so far was a one-man incarnation of how the whole field operated: pick a piece of the universe, study it until you hit diminishing returns, then put it down and pick up another one. The thing itself, the coherent universe, seemed to be lying in fragments all over the place.

Jim didn't think he'd been wasting his time, but he suspected he was casting about, looking for something to get passionate about, something with enough intellectual heft to be irresistible.

———

In early 1987, Kitt Peak National Observatory had a mirror it had commissioned as an experiment and now didn't know what to do with. To brainstorm options, it held a meeting in Tucson on February 18, 19, and 20, 1987, and Jim went. On the first day, Jim gave a talk in which he got on his high horse about understanding the universe from the galaxies up and said that the most exciting use for a telescope with the Kitt Peak mirror was something that had been "simply impossible before, namely the direct study of the evolution of galaxies."

Then other astronomers gave talks outlining other ideas, and while Jim sat there listening, he stopped thinking about galaxies and thought about technology instead. The Kitt Peak mirror might be a good match for the newest CCDs. Just recently Jim had been at Caltech and run into an engineer named Morley Blouke who had opened a polyethylene pizza box and showed Jim the latest CCD. WFPC's state-of-the-art CCDs, which had been made by Blouke's company, were 800 pixels on a side, 6,400 pixels altogether, meaning a field of view of 10 arc minutes, a third of a moon, across. The CCD in the pizza box had 2,048 pixels on a side, 4 million pixels altogether. Looking at those enormous CCDs, Jim felt they were the wave of some future he wanted to be part of.

Jim, listening in Tucson, thought he could arrange the new CCDs into a mosaic and build a camera with a field of view of 120 arc minutes, that is, 2 degrees or four moons across. Moreover, Jim thought, he could use the same CCDs in a spectrograph, and if the CCDs' pixels were fed by the new optical fibers that happened to be a good match to the size of the individual pixel, the spectrograph could take the spectra of hundreds of galaxies at once. The spectrum of a galaxy shows much more than its redshift/velocity/distance. The amount of light an object has at the different wavelengths also shows its inner workings, its physics: what elements it's made of, how fast it's rotating, how hot it is, and how old it is. Images and spectra together tell you everything about an object that, from a distance, is possible to know.

He could put this camera and this spectrograph on a telescope with the extra Kitt Peak mirror and let the telescope drift scan along one strip of the sky after another. All those strips could be stitched together into a survey, which if run for five years would give a highly informative three-dimensional picture of a large fraction of the nearby universe. It could take the spectra of hundreds of thousands of galaxies—the same survey with current spectrographs would take several thousand years—and the images of tens of millions of galaxies. It would specify and track the whole galaxy zoo. It would be digital: images and spectra of wide fields of the sky, debugged and cleaned up and analyzed and dumped into your computer.

Jim talked about his idea in the meeting's hallways and over dinner, and by the end of the meeting, everyone had discussed it. They saw no big technical problems with it and in fact seemed excited. When asked whether they'd be willing to work on a proposal to build what they were now calling the digital telescope, three-quarters of them raised their hands.

After the meeting, Jim worked out that the mirror that most suited the size of the CCDs and the size of the optical fibers was smaller than Kitt Peak's. A little 2.5-meter mirror, he said, would be about right. Then he went back to the Hubble Space Telescope and WFPC and the quasar and cluster surveys and dropped the subject. The talks given at the Tucson meeting were written up and printed out and bound with a nice plastic spiral, and then everybody else dropped the subject too.

Chapter 2

Chicagoland

It was obvious for a long time that some large homogeneous monolithic thing needed to get done, to get to where the rubber hits the road.

—Jim Gunn, Princeton University

AN ASTRONOMICAL SURVEY is not only a map of where things are, it's also a census of the kinds of things out there. The impulse to look at the sky and re-create it in cave paintings, clay tablets, papyrus, silk, parchment, paper, glass plates, and photographic film—has been so universal that it must be hardwired. The earliest systematic survey, done in the late 1700s, was by William Herschel, who built his own telescopes and moved them methodically across the sky, up and back, making what he called "sweeps." He found a new planet and drew an amoeboid-looking map of the Milky Way, and though he and everyone else still thought the Milky Way was the whole universe, he also saw what he thought were nebulae, "cloudy things," which much later turned out to be other galaxies. He listed all objects, their positions, brightnesses, sizes, and possible classifications, and annotated his list with comments: what's now called the Whirlpool galaxy, he wrote, was "surrounded with a beautiful glory of milky nebulosity." His list later grew into the New General Catalogue, and thousands of objects are now identified by NGC numbers. Herschel said that the power of a survey was in statistics.

This idea doesn't sound lovely, but it is. It means that you can understand an object's life either by watching it from birth to death or, if its life cycle is longer than yours or even humankind's, by looking at so many examples of that object that you see all its life stages. Herschel said this more poetically: "[The heavens] are now seen to resemble a luxuriant garden, [and] . . . is it not the same thing, whether we live successively to witness the germination, blooming, foliage, fecundity, fading, withering, and conception of a plant, or whether a vast number of specimens, selected from every stage through which the plant passes in the course of its existence be brought at once to our view?" Fred Hoyle said it more simply: "The Universe is so vast, and the lengths of time . . . are so long, that almost every conceivable type of astronomical process is still going on somewhere or other."

All these surveys were done in the visible, in the optical wavelengths in which we see. In the mid-1900s, the first nonoptical survey was done by Cambridge University in radio wavelengths. For the first time, we saw objects giving off radio waves—among them were quasars. This and subsequent radio surveys were gathered into three Cambridge catalogs and the objects in them identified with 1C, 2C, or 3C numbers; Maarten Schmidt's first quasar, for example, was 3C 273. The objects in the Cambridge catalogs changed with distance. That change, along with Edwin Hubble's graph showing the universe expanding, implied that the early universe was different than the one nearby, that not only do stars have life cycles, but so does the universe. After that, every time astronomers invented an instrument that could detect other wavelengths—gamma rays, X-rays, ultraviolet, infrared, microwaves—they did a survey of the sky and found things they hadn't known were out there.

Surveys like Herschel's were done by eye and recorded by hand, so the first optical survey done with a camera, the Palomar Observatory Sky Survey, became the gold standard of twentieth-century astronomy. Done with Caltech's 48-inch telescope on Mount Palomar and begun in 1949, it covered the sky over the entire Northern Hemisphere, took ten years, and produced four thousand glass photographic plates. Astronomers could buy photographic prints of the plates for $14,000 or glass copies of them for $25,000; most astronomy departments bought one or the other. When astronomers like Maarten Schmidt found something weird in a radio survey or an X-ray survey, they would go to the Palomar plates to see what it was. Look through an optician's magnifier at a little smudge on a glass plate, and the smudge turns into a delicate, perfect galaxy. Lay a universally agreed-on grid over the plate and note the galaxy's coordinates, and then go to your own telescope, set it to the same coordinates, take the galaxy's spectrum, and learn all about it. The Palomar survey was not only beautiful, it was a mine of information.

It was also inherently flawed. The resolution, the detail on the plates, was exquisite. But the plates could give little idea of how bright any given object was. The glass plates were coated with grains of silver halide that darken when exposed to light, and twice as much darkening doesn't necessarily mean twice as much light. For a certain amount of light, the plates respond linearly; that is, the amount of light and the darkening track each other. A lot more light, and the plate gives up and saturates—it stays the same deep black no matter how much more light hits it. You couldn't measure an object's brightness with an accuracy of better than 10 percent. Moreover, the Palomar plates, like all photographic plates, differed from one another. Plates were exposed at different times, under different conditions. So you wouldn't know whether stars measured at twelfth magnitude on one plate were the same brightness as twelfth magnitude stars on another plate. Jim Gunn used to say that astrophotography was a black and careful art done by people who were masters, all of whom seemed to be much older than he was. This was why, after centuries of hard work, cosmology was still asking basic questions and still unable to believe its own answers.

So CCDs were a godsend. Light is measured in photons, and almost every photon hitting a CCD from a galaxy triggers an electron out to a computer. Glass photographic plates recorded maybe 5 photons in 1,000, about a half of a percent of the incoming photons. CCDs record 700 in 1,000, 70 percent. CCDs register immediately, respond linearly to light, and don't saturate until hit with 20,000 times more light. With plates, an object's brightness can be measured to no better than 10 percent; with CCDs, to about 0.5 percent. A twelfth magnitude star on one CCD is a twelfth magnitude star on the next CCD. The biggest problem CCDs had in the late 1980s was that the amount of data coming off them overwhelmed the computers: one image from one of the new CCDs was 8 megabytes of data, the size of a large book.

Assuming the data problem could be solved, a CCD survey of a large fraction of the sky, being digital, could be sent straight from the telescope to a computer, and from there could be downloaded to computers everywhere. A digital survey would record everything we could learn from images and spectra: it would have the brightness, size, color, and shape, the composition, rotation, temperature, age, velocity, and distance of every one of an enormous number of galaxies. It would be standardized and evenly calibrated—a computer applies the same corrections for, say, dust to all observations. It would live in computers. It would be portable and easily accessible and nobody had to master the black arts to use it.

Rich Kron, a cosmologist at the University of Chicago, hadn't been at the Tucson meeting where Jim Gunn first talked about a digital survey, but he had read the spiral-bound write-up of it. He was a little surprised that the meeting reached such consensus —unusual for astronomers—and that nothing had yet come of it. He thought the idea was too good to drop, and in fact he'd had a similar one himself, though he had doubts about it.

Rich is a kindly man, modest and serious; even his smiles are serious. His dark hair is cut straight across his forehead, and you suspect he must have looked much the same as a child. Rich's mother and father were both astronomers who lived at the Lick Observatory, just east of San Jose on the top of Mount Hamilton. Mount Hamilton is actually a steep, thin ridge, and without trying hard you could actually fall off, a possibility that was Rich's idea of bliss. The observatory employed about one hundred people, which qualified it for a one-room school, and on Rich's way to school, he'd bicycle past the second-largest telescope in the world. Rich went into a youthful astronomy stage and never came out.

After high school, Rich spent a year at a boarding school in Marlborough, England, just north of Stonehenge, because the school had a 10-inch refracting telescope—a larger version of the kind Galileo had used to discover that the Milky Way was full of stars—that was out on the downs, surrounded by Neolithic tumuli. Rich and a few other diehards went out to observe regardless of the weather, which was usually bad. When it was, they'd just have coffee and talk. Rich felt it was one of the best years of his life.

After college, he went to graduate school at the University of

California, Berkeley, which had one of the country's best astronomy departments. He felt he was a member of the department, given responsibilities and allowed to get his hands on something real. His dissertation was on whether a galaxy's color could be used as a proxy for its redshift. At the time, the mid-1970s, getting a redshift required hours of taking spectra, but measuring color just required taking a picture. His plan was to photograph a whole field of galaxies and plot them on a graph, faint to bright against red to blue. Then he would follow up the photographs with spectra to find their actual redshifts. He expected that the faintest galaxies would have the highest redshifts and look the reddest and the brightest galaxies would have the lowest redshifts and look bluest. He was granted an astoundingly generous forty nights to observe at Kitt Peak—most astronomers get maybe four nights. He began by photographing two patches of the sky, each with both a red filter and a blue filter, and then developed the plates. With the red filter, the red galaxies popped out from the background, and he found what he expected: the faint red galaxies. But on the plate with the blue filter, he found a scientist's delight: the wholly unexpected. The blue plate was covered with faint blue galaxies, a completely unknown population, as though someone had surveyed a town and missed all the adolescents. Holy cow, he said to himself, look at all those galaxies!

After he finished his PhD in 1978, he was hired immediately at the University of Chicago and its allied Yerkes Observatory. Like most cosmologists of the time, he was entranced with the early universe, with finding the next most distant thing, then the thing beyond that. At Chicago, he got right to work. He'd locate a random, dark-as-possible patch of the sky, take the spectra of everything in it, and study it all to death. He was looking for anything—quasars, stars, clusters of galaxies—but in particular for galaxies just forming. No one had a good notion of what such galaxies would look like, and he didn't find anything he'd nominate as a candidate. He did find that galaxies farther away were intrinsically much bluer, and he wondered if maybe those blue galaxies could be babies. Eventually he and several colleagues used Kitt Peak to map four hundred galaxies out to a redshift of 0.1 in tiny patches of sky, narrow surveys that they called pencil beams. In the pencil beams, Rich and his colleagues found, at regular intervals, peaks in the numbers of galaxies every 100 megaparsecs or 326 million light-years, as though their pencil beams were running through walls of galaxies, one wall after another. They thought they were seeing a sort of cross section of the universe's network of networks, its structure on the large scale.

Start with the universe on the small scale and go to the large. Earth is a planet, held by gravity—along with seven other planets and lots of smaller entities—in orbit around our star, the sun. This solar system is itself held by gravity—along with 10 billion other stars and solar systems—in an orbit around the center of a barred spiral galaxy. We see the galaxy up close and edge on as a splash across the night sky and call it the Milky Way. The Milky Way is bound by gravity—along with around thirty-five galaxies, including Andromeda and the Large and Small Magellanic clouds—in a cluster; ours is called the Local Group. The Local Group is one of around one hundred galaxy clusters gathered by gravity but not bound together and strung like beads into the Virgo Supercluster. The Virgo Supercluster is one of an uncounted, uncountable number of superclusters that in turn form what looks like an elaborate, fractal foam of galaxies arching like the skins of bubbles around immense voids.

This whole shebang, this large-scale structure, must have formed itself out of the ancient nearly featureless haze of matter left over from the Big Bang. And it must have been formed by gravity, because no other force operates on that scale. But as of the late 1980s, no one knew how. Maybe it formed from the bottom up: galaxies formed first and then collected into clusters, which collected into superclusters. Maybe it formed from the top down: superclusters fragmented into clusters, which fragmented again into galaxies. In fact, except for a few superclusters and some voids with no galaxies at all, no one knew what the universe's large-scale structure looked like or how big it got. Observers painstakingly mapped superclusters, galaxy by galaxy and redshift by redshift, until the superclusters reached the edges of the maps. And no matter how much bigger the maps got, the large-scale structure still ran off the edges. No one knew the limits of what we could see, no one had a complete map of the universe. Cosmologists seemed a little silly, worrying about how the universe would end, when they didn't know what it even looked like.

Rich thought this problem of large-scale structure was exceptionally interesting. In April of 1988, before he'd read the spiral-bound write-up of the Tucson meeting, he was on his way out to Lawrence, Kansas, for a workshop on large-scale structure. On the airplane, he thought about an idea he'd had for mapping the large-scale structure by doing a redshift survey on a large fraction of the sky. What might he be able to do, he wondered, if he had a telescope that he didn't have to share, that would be dedicated only to such a survey? He didn't know much about CCDs, let alone about the new CCDs in Blouke's pizza box. But he imagined a survey done with a telescope that had both a camera with a lot of CCDs and a spectrograph whose CCDs were fed with a lot of fibers, allowing the spectrograph to take a lot of redshifts simultaneously. Such a survey could end up with both images and redshifts of a million galaxies, and in digital form. He wasn't terrifically serious about this because he was not an instrument builder—he thought of himself as just a dumb observer—and would have no way of knowing whether such a thing could actually be built or whether it would work. When he got to the Kansas workshop, he gave a talk about his idea called "Redshift Survey of Faint Field Galaxies," but feeling modest, never put the talk into writing. When he finally read the Tucson write-up and Jim Gunn's idea for a digital survey, he thought, "Darn it. It's so exciting, somebody's got to do it."

As it happened, Rich and Jim had been meeting regularly, usually by phone, with Don York, an astronomer who was Rich's colleague at the University of Chicago and who had been at Princeton with Jim about five years before. At Princeton Don had worked with Lyman Spitzer, who first thought up the Hubble Space Telescope, on a less sexy space telescope called Copernicus. Once Copernicus launched, Don and his colleagues used it to find evidence of an element, deuterium, in such quantities in the universe that cosmologists—Gott, Gunn, Schramm, and Tinsley included—changed their estimates of the universe's density. By that time, 1982, Don had moved to the University of Chicago as a professor of astronomy; a couple of years later, he would become director of the Apache Point Observatory on Sacramento Peak in New Mexico. Apache Point was in the process of getting a telescope with a 3.5-meter mirror, and Jim, Rich, and Don, all friends, had for years been discussing the telescope's slow progress. Now they began discussing a digital sky survey and got charged up about it. They decided that if they got some more people together to discuss it, maybe somehow something would happen.

Getting people together to discuss things was unusual. Astronomers normally work in small groups of one or two or three and think of themselves as lone explorers. Doing research with a large team was the last thing Rich wanted to think about, and Jim felt the same. When Jim and a colleague—Maarten Schmidt, Bev Oke—thought of something new to build, they just found the money and built it. But a digital sky survey seemed as though it might cost a lot of money, and certainly the CCDs would, so in addition to getting other people as charged up as they were, Jim, Rich, and Don needed to think about where to get money. The National Science Foundation (NSF) was the only government source for funding astronomy done from the ground—NASA funded astronomy done in space. Since the NSF was more likely to give large sums of money to large collaborations than to small groups, the obvious route was to find more collaborators.

They wrote a polite and enthusiastic letter to their fellow cosmologists, who might be either collaborators or competitors. Writing to all these people, though unusual, was politic. Astronomers operate in a hair-raisingly closed system: any dollar one of them gets is a dollar someone else won't get; everyone knows everyone else; and the funding agencies ask them to judge one another's grant proposals. In such a system, the best policy is to be open and to be sure that your project won't duplicate anyone else's, that all research niches are separate but complementary.

Soon after, on September 6 and 7, 1988, fourteen astronomers arrived at Chicago's O'Hare Airport, met in a conference room in the Hilton hotel on the airport's second floor, had dinner downstairs in a restaurant in the hotel lobby, and never had to leave the grounds. They agreed immediately on mapping out the large-scale structure of the universe over a substantial fraction of the sky. They considered many questions: Could they make the map by patching together all redshift surveys currently under way? Or would a single survey done to a single standard be more scientifically credible? Should they build their own telescope dedicated to the survey or try to usurp an existing telescope by making an offer its owners couldn't refuse? Jim described the telescope, spectrographs, and camera he'd already described in Tucson and more or less stopped the questions cold: it was so obviously the way to do it.

They gave themselves assignments for further research, and the following December they had another meeting, again at O'Hare. They figured the spectrographs would cost a little over $1 million, the camera would cost $2.7 million, and no one knew how expensive the big CCDs would be. They wondered whether they could afford to archive all that data and decided that when the time came, they could handle it. They agreed to site the telescope either at Kitt Peak or Apache Point, and they thought they might call the project "A Complete Survey of Galaxies." Rich with his straight-across haircut and earnest smile said, "The grand and bold thing is what we want to do."

The NSF, however, didn't want to fund the survey. NASA didn't seem to be a good match, and the Department of Energy seemed even worse. The alternative to finding one big funder was to start smaller, with individual universities. No one thought universities would be able to fund the whole survey, but smallish university grants can be important in the deliberations of larger private foundations or government entities as votes on the likelihood of the survey being smart and successful. If your own university doesn't believe in you enough to give you money, who else would?

Rich Kron's home base, the University of Chicago, has a reputation for high-minded academics with applications to hard reality. Its departments of sociology, education, and economics have fundamentally changed cities, schools, and national fiscal policy; its physics department hosted the proof of principle for the atomic bomb. The university is an island of English Gothic—stone spires, flying buttresses, gargoyles, and all—in Chicago's Hyde Park, itself an island in the middle of the rough South Side. Princeton and Caltech are set in historic and charmingly preserved towns; Chicago's South Side has inner-city high-rise housing projects and ethnic neighborhoods with a history of racial tension. Wander out of one of the University of Chicago's landscaped quadrangles through a flame-shaped gate without looking, and you'll get smacked flat by traffic or maybe even shot.

Rich and his colleague, Don York, hadn't gone to the first O'Hare meeting expecting to ask the University of Chicago to support the survey. Chicago already had a share in a telescope, the 3.5-meter at Apache Point, and Chicago astronomers already had other irons in the fire, large potential projects. But between the first and second O'Hare meetings, the projects fell through. Rich and Don felt ridiculous going to the provost and saying yes, they did ask for big projects, but now they'd changed their minds and wanted a small telescope and a survey. So they weren't even going to ask. But the department of astronomy's chair, Don Lamb, was now worried about his department's future in optical astronomy, and that fall, he met with the provost. The provost agreed to the survey immediately and allowed them $34,000 to buy a place in line for the creation of a 2.5-meter telescope mirror.

A few months later, Don York made the first serious budget: not including salaries for scientists, who would surely donate their time, the total cost would be $12 million, including $558,000 for the mirror, which could be resold if the survey didn't launch. No one still knew what the CCDs would cost, just a lot and maybe a sizable fraction of the cost of the whole survey. The Chicago astronomers began fund-raising and between the 1988 O'Hare meetings and 1990 met with a complete lack of success.

Meanwhile, the department's theorists were in the middle of creating a new field called particle astrophysics. It was a marriage between high-energy physicists looking for the rules governing the universe's fundamental particles and cosmologists studying the young universe—the connection being that the young universe was the last time those fundamental particles were not bound up into atoms and were still at large. Toward this end, two Chicago theorists, Michael Turner and Edward "Rocky" Kolb, had helped set up the Theoretical Astrophysics Group at the nearby Fermi National Accelerator Laboratory, a large physics lab and particle collider run by the Department of Energy. A third Chicago theorist, David Schramm—of Gott, Gunn, Schramm, and Tinsley—was on Fermilab's Board of Overseers. Michael Turner was also on a Fermilab panel to decide the direction it should take under its new director, John Peoples. Given all these mutual interests and feeling desperate about funding, Chicago's Don Lamb thought he'd meet Peoples and see whether a government laboratory dedicated to physics might be interested in a university survey by astronomers.

After a high-level dinner party and a lot of talking, the survey and Fermilab turned out to be a good match. What Fermilab could do for the survey was simple: the survey was going to produce more data than astronomers had ever handled before, and Fermilab physicists handled such amounts routinely. What the survey could do for Fermilab was more complicated: the world in which Fermilab operated made academia look sheltered and sweet.

Chicagoland is the city of Chicago plus 7,000 square miles of rich Illinois prairie long since paved over with the national cliché—four-lane highways, strip malls, parking lots, chain stores, and subdivisions that reach the horizon. Drive through thick, aggressive traffic down Interstate 88, turn right on Kirk Road, turn right again onto Pine Street, show the Fermilab security guard your driver's license, and suddenly you're in a presuburban century, on the prairie that the glaciers planed off flat, checked with a level, then planed again and again—10 square miles of dead-flat land, prairie grasses, and air. In every direction, the sky curves down to the ground uninterrupted. You can stretch your arms out on either side as far as you can reach and never hit another thing; you can get your breath.

Growing out of this prairie is Fermilab's headquarters, a building made of glass whose two sides curve in toward each other, meet, and rise for fifteen stories. Inside the building, the offices either look out onto prairie or in on the atrium. The atrium goes straight up fifteen floors to the ceiling, and the ceiling is glass. You stand in the atrium and look up fifteen stories, then through the ceiling, and your eye beams don't stop until they hit infinity. It's the perfect place to do high-energy physics: to accelerate fundamental particles to near the speed of light, smack them into one another, and from the debris, study the physics of the universe's fundamental forces.

But Fermilab is also a government laboratory, run by the Department of Energy (DOE), and in 1989, when John Peoples became director, DOE had just decided that Fermilab would not be the site of the world's largest new physics experiment, called the Superconducting Super Collider, known as the SSC. High-energy physics experiments are so big and expensive they are built only one at a time, and DOE had decided the SSC would be built in Texas. So Peoples' first job in taking over Fermilab was to figure out how to avoid presiding over its demise.

One way was to create a new mission for Fermilab to form a complement to the Theoretical Astrophysics Group and call it the Experimental Astrophysics Group. The name was an artful fudge so Peoples could avoid a cultural firestorm at home: astronomers are not experimentalists, but physicists are. Peoples asked Rich to take a joint appointment at Fermilab, teach astronomy to some physicists, and head the Experimental Astrophysics Group, which could then become a wedge for joining the digital sky survey. Peoples thought that a new project would give Congress and DOE less reason to shut Fermilab down. Besides, Peoples had a yearly budget of more than $200 million, and once he understood the survey's size and likely cost—even though he didn't yet understand entirely what surveys were good for—he thought, "Jesus, even if they're off by a factor of three, this doesn't cost anything!"

Peoples ran into physicists' objections at home for a while but arranged enough talks and meetings and damped enough fires that by the spring of 1990, he could agree officially to commit to the survey.

In the summer of 1991, Rich was at a scientific conference on quasars, at which Jim Gunn's old student and a fellow quasar hunter Don Schneider was giving a talk on the survey. Don showed some transparencies, described the camera, and said that when they finished the survey, they'd have the spectra of 100,000 quasars. At the time, the quasar hunters had collected around 5,000 quasar spectra, so Don was claiming that the new survey would find twenty times the number of quasars found in the previous thirty years. Don had a reputation as a joker, so the audience laughed. Someone yelled out, "Good try, Don." Rich had to stand up and explain that it wasn't a joke.

Chapter 3

Putting on the Play

For every dollar you put in, Princeton matches it. And for
every dollar that Princeton puts in, Chicago matches it. And
then for every dollar the universities put in, NSF matches it.
And for every dollar NSF puts in, Sloan matches it. OK? The
multiplicative factors add up.

—Jerry Ostriker, Princeton University

God, they've been doing this since Isaac Newton,
can't they get it right?

—Hirsh Cohen, Sloan Foundation

ABOUT THE TIME the University of Chicago was founded,
Princeton University turned 150 years old. Nassau Hall,
its main building, had been the venue for the Continen-
tal Congress in 1783, just after the university amended its charter
so that its trustees no longer needed to swear allegiance to the
king of England. The gray stone Gothic campus still feels a little high-church royalist, lined with long windowed cloisters and
arched doorways that look onto wide courtyards and sheltering

trees. Walk through campus and you feel protected, surrounded by privilege, order, beauty, and authority. Downhill and off on the campus margins is the Department of Astrophysical Sciences' Peyton Hall, which looks like a sunken double-wide trailer. Peyton nevertheless houses a splendid department: of the seven Crafoord Prizes—the equivalent of Nobel Prizes for astronomy—given so far, Princeton astronomers, including Jim Gunn, have won three.

After Jim got back to Princeton from the O'Hare meetings, he went looking for his department chairman, Jeremiah Ostriker, who Jim thought was a mastermind at finding money. Twenty years earlier, when Jim was in his first brief stint at Princeton, they'd collaborated on a theory explaining pulsars, and they'd been friends ever since. Ostriker, universally called Jerry, is a vivid, jumpy, short astronomer who talks right at you with a fast New York accent. He is a theorist who works on understanding nearby objects and their evolution in terms of the laws of physics; only early in his career did he ever go near a telescope. "I actually observed in person," he said. He also belongs to a small subset of astronomers, referred to as astropoliticians, who are good at building the next node in the next socio-politico-scientific network and at influencing the intentions of universities and donors and government agencies.

Jerry and Jim had a brief meeting: Jerry just asked if Jim was sure he could build the instruments. Jim said that all the pieces were pretty much available off the shelf, and yes, he was sure. Jerry was convinced—"The wonderful thing about Jim's instruments is that they work," he said, "and this is not true of everyone's instruments." So Jerry told Jim, "You build it, I'll find the funding."

In a miracle of timing, two weeks later Jerry got a phone call about a Princeton graduate, Keith Gollust, who was interested in funding science projects. Jerry met Gollust and outlined a plan for obtaining matching dollars from other universities, agencies, and foundations: "So it's pretty much guaranteed that every dollar from you is going to be worth several times two," Jerry said. "You know, two times two times two times two is probably worth sixteen." Gollust apparently agreed with Jerry's multiplicative thinking and agreed to give the survey money, probably around $2 million.

Then Jerry enlisted the support of Princeton's own heavyweight neighbor, the Institute for Advanced Study, famous for having been the American home of Albert Einstein. Around Princeton, it's just called the Institute. Compared to Princeton's stone courtyards, the Institute gives the impression of having been built and maintained to budget, but the trees are huge, the surrounding meadows are mowed and conducive to walking, and for a place in the middle of the I-95 corridor, it's deeply peaceful. Inside, its hallways open up into informal meeting places with chairs, tables, and blackboards, and the person in the next-door office is probably the world record holder of some feat of intellect. It has only a handful of eminent faculty, holds no classes, and hosts a rotating roster of visitors who stay anywhere from a few months to a few years. Its faculty and visitors don't teach, they don't experiment, they don't do observations; they just think and talk and write.

The only astronomer on the Institute faculty was the same John Bahcall who had invited Jim to the Chinese diner to talk about the Hubble Space Telescope. Jerry, Jim, and Bahcall were old friends, having collaborated in every possible combination of pairs for decades. At the Institute, Bahcall ran what Jim called "an incredible postdoc shop." Bahcall, who had good taste in postdoctoral fellows, offered them jobs that were then unique: they'd be paid to work on whatever they wanted to. Bahcall dropped in on the postdocs regularly, helping them think through their work whether or not their research was in his field, lunching with them argumentatively, looking out for jobs for them. All told, the Institute postdocs had the most prestigious and desirable fellowship a young astronomer could hope for. Because the Institute was in the business of thinking and not of observing, it certainly would not want to be part owner of a telescope. But Bahcall liked the idea of a survey and trusted Jim, so he told Jerry he'd like to join the survey provided he didn't have to go to meetings. The Institute promised a nominal amount of money to the survey, in return for rights to use the data, and Jerry had his eye on those Institute postdocs.

Next, Jerry did something unusual. He wrote a draft of what he called the Principles of Operation—and came to be known as the PoO—that all but gave away the farm. The data from the survey would, if the survey ever managed to get money from the NSF, by mandate have to go into a public archive. But the survey members could still have a year when the data was proprietary—time during which the cream could be skimmed, discoveries and careers made, highly cited papers written. The PoO stipulated that during the proprietary year, the survey's data should be open immediately not only to the people working on the survey but to anyone at any member institution. Furthermore, no one had dibs on any subject; anyone wanting to work on a specific subject would notify the whole team first and anyone else wanting to work on that subject would sign on, and whatever fame resulted would have all their names on it.

Some people who weren't Jerry worried that some astronomer in one of their departments would be smart enough to sit back, watch everyone else do the work, and then swoop in and snag the big discoveries. Jerry told the worriers, "Look, you're designing it, you'll have more knowledge of it all than anybody else, don't worry about it." He thought the "mine, mine, mine syndrome" was an inefficient use of the sky. Everyone discussed the issues extensively and eventually agreed on the PoO. The survey would have enough science to go around; they could afford to share. They were a little surprised at themselves, though. No other astronomical project had ever been run this communally.

In April 1990, the presidents of the three institutions—the Institute's Murph Goldberger, Princeton's Harold Shapiro, and Chicago's Hanna Gray—put in $350,000 each and expressed hopes for the future. In September 1990, Murph Goldberger announced that a donor, Richard Black, was "clearly and unmistakably seriously interested in the Survey Telescope Project." And then in November, Fermilab decided to join, adding its hardware and software capabilities.

So at the end of 1990, partners in place and money promised, the survey did what astronomers have always done well: they notified the press. The *New York Times* wrote that the largest array of light sensors ever assembled would find bright enigmatic objects deep in space, solve great cosmic puzzles, and create an elegant and detailed map of the sky. The survey hadn't quite settled on a name for itself; it had been called variously the Sky Survey Project, the Survey Telescope Project, the 2.5-Meter Survey Telescope, the Million Redshift Survey, the Map of the Universe, and the latest name, A Digital Sky Survey of the Northern Galactic Cap. The Northern Galactic Cap is what it sounds like, the top part of the sky over the Milky Way's north pole, a quarter of the whole sky; it's a precise name, but not catchy.

To help with fund-raising, Rich Kron had put together a brochure called *Map of the Universe*, whose front cover was a hand-drawn, edge-on Milky Way, looking like a tadpole with a tail at each end. The brochure explained that the survey would solve the great outstanding problem of structure in the universe by creating a map of the sky hundreds of times larger than any other map.

Don York had been taking on the survey's management-type jobs and for the third time, put together an official budget: "These costs look a little frightening," he wrote, "but they are real." The cost had gone from $12 to $14 million. The CCDs that were the heart of the survey finally had a price: nearly $2 million. The polished mirror would be nearly $1 million. And the software to run the instruments and the survey would be $279,000, and this didn't include the seven man-years of time to write it. "Obviously there is much to be saved," Don wrote, "if people with full-time university salaries take on responsibilities at no cost." The commitment would be enormous, not only money but five years of everybody's lives before any payoff whatever.

By late summer of 1991, the cost of the survey had bounced to $25 million. So Jim, Don York, and Jerry paid a visit to the Sloan Foundation—the Institute's Murph Goldberger was on its board—at its fancy address in New York City. They went upstairs to the wood-paneled boardroom, got out their viewgraphs, and explained the survey to the foundation's director, Ralph Gomory, and the program officer for science, Hirsh Cohen. Jerry thought Gomory and Cohen looked interested, Jim thought they were impressed, and Don thought it seemed pretty clear that the astronomers were in.

The astronomers were a little too sanguine. Hirsh Cohen was worried because the survey needed a lot more money than the few hundred thousand dollar grants the Sloan Foundation usually gave. Furthermore, the foundation's stated mission was funding science with "a high expected return to society," and Hirsh suspected that any returns to society from an astronomical survey would be at a distance measurable in kiloparsecs. But Hirsh was bowled over that the three university presidents had antied up cash—he'd once been an academic and knew that sort of thing rarely happened. And Jim and Don and Jerry had been good at pitching, Hirsh thought; they were very serious guys. So he and Gomory asked them to send in a proposal.

The foundation took a few weeks to get outside reviews, first doing what Hirsh thought of as snooping around, talking informally to unaffiliated astronomers, then asking them more formally for written recommendations. What sold the foundation was that the data would be free and held in a long-lasting archive open to the world—surely, a high return to society. Within a few months, the Sloan Foundation offered the survey an unprecedented $8 million, provided that any budget overruns would be paid by the universities. The universities agreed, and the survey in gratitude named itself the Sloan Digital Sky Survey, or for short the Sloan. After a while they called themselves Sloanies.

But the $8 million promised by the Sloan Foundation, along with the $1-million-plus promised jointly by Princeton, Chicago, and the Institute and the $5 million promised by Fermilab, weren't adding up to the $25 million needed.

In the fall of 1991 the president of Johns Hopkins University, Bill Richardson, notified the Hopkins astronomers that the university would no longer participate in building a giant 8.5-meter telescope called Magellan. Over the past few years, ever since the Space Telescope Science Institute had moved across the street, Hopkins had been beefing up its contingent of astronomers. The problem was that luring hotshot astronomers to your faculty is hard unless you have a telescope, and, like Princeton, Hopkins didn't. So Hopkins had bought a share in the Magellan. But Magellan was now running late, and Hopkins's share was going to climb to around $10 million, and President Richardson didn't see how he was going to raise that kind of money, so he'd told the Hopkins astronomers that he was sorry but he was pulling them out of Magellan. He invited them to come up with another project that would keep the department beefed up but that would cost a lot less.

The astronomers began looking around for alternative telescopes. At the same time, their department happened to be holding a departmental review, a routine review by outsiders of a department's flaws and virtues. On the review panel, by chance, was Jerry Ostriker. Jerry knew that Hopkins had a long tradition of spectroscopy and that the Sloan survey's spectrographs were to be built by engineers at the University of Washington who already had a lot on their plate. He also knew that President Richardson had moved to Hopkins from the University of Washington and had been a graduate of the University of Chicago and knew Hanna Gray. The stars thus aligned, Jerry said to the Hopkins astronomers, "Why do something dumb like buy a piece of some random telescope? We're going to do this whole new sky survey. It's all exciting." One of the Hopkins astronomers, Tim Heckman, thought that Jerry could sell the pope a double bed.

Hopkins was pleased with the prospect of providing the spectrographs for the project—all it had been able to do for the Magellan was send checks. But what most impressed Heckman was what had impressed Hirsh Cohen: the potential contents of the archive, that is, the number of different astronomical fields covered by the Sloan survey and the amount of data in all the fields. Just reading the proposal, Heckman could see that the Sloan was going to change the way astronomers without private telescopes worked: instead of taking a couple of years writing and rewriting proposals for three clouded-out nights on a public telescope, you could just take the interesting question that occurred to you and find the data you needed in the library of the universe. And the interesting questions would not be just about large-scale structure: for every question in optical astronomy that Heckman could think of, the Sloan archive would have data. And the amount of data on each question would be orders of magnitude larger than it had ever been before. Heckman talked to his dean and provost, who talked to President Richardson, who thought the survey was a great idea and that the worst-case cost to Hopkins of $2 million was better than Magellan's $10 million. In June 1992, Hopkins signed on officially and agreed to build the Sloan survey's spectrographs.

Hopkins's tradition of spectroscopy had started with the nineteenth-century physicist Henry A. Rowland, who had figured out how to make spectrographs with near-perfect diffraction gratings (alternatives to a prism) that spread light into a rainbow much the way compact discs do. Over the next century Hopkins had built on the excellence of Rowland's diffraction gratings by hiring physicists and astronomers who relied on spectroscopy and engineers who were expert at building spectrographs.

Jim had done a design for the Sloan spectrographs—two of them, to take advantage of the telescope's large field of view. Light would come in from 640 quasars or stars or galaxies, be passed on through 640 optical fibers, sent to a combination of a diffraction grating and a prism—called a grism—which could fan the light out into 640 little spectral rainbows, which would register on the same-sized CCDs as in Jim's camera. At Hopkins, Jim's design was taken over by Alan Uomoto, an astronomer/spectroscopist who welcomed building the spectrographs—he thought it would be mostly a matter getting the parts and putting them together. He was prepared to work full time on the spectrographs for the next two years. A few months later, he amended that to three.

Uomoto hired a young man named Steve Smee, who had an undergraduate degree in mechanical engineering, had started out as a machinist, worked on optics a little, and was by all accounts brilliant. Uomoto said his own brilliance was in hiring Steve Smee. Uomoto and Smee worked together the way astronomers work with engineers, astronomical idealism balancing engineering reality; if the engineers can convince the astronomers to relax the instrument's precision by 50 percent, it'll cost only a tenth as much to build.

Originally, the spectrographs were to have been built not by Hopkins, but by the engineers contracted out to the astronomers at the University of Washington. Astronomy departments often have small dependent cadres of engineers who help design and build their instruments. Paid with the astronomers' grant monies, they work until the grant runs out and have little or no job security. Oddly, they often aren't trained as engineers but have fallen into their jobs because they're logical, analytical, clever, and have what scientists call "good hands."

The engineers at the University of Washington—which calls itself UDub, short for UW—were chosen because they had been building the 3.5-meter telescope at Apache Point, owned by the Astrophysical Research Consortium (ARC), a group of universities to which Princeton, Chicago, and UW belonged. ARC was now winding down the construction of the 3.5-meter and was therefore about to stop funding the UW engineers, but Bruce Margon, ARC's director and chair of UW's astronomy department, wanted to keep the engineers around—he thought they were one-of-a-kind people. So the arrangement was that the UW engineers would provide the Sloan's hardware: they'd subcontract out the mirrors but they'd build the telescope, its enclosure, and the spectrographs. No one—not Jim, Don, or the engineers themselves—doubted their ability to build all that hardware. But they were a small group and it was a lot to take on, and it all had to be done on time. They were annoyed when Don told them he'd given away the spectrographs, but he explained that Hopkins wouldn't join unless it had something to do.

Before making the 3.5-meter, none of the UW engineers
had built a telescope nor, true to form, were they even trained
as engineers. Walt Siegmund, for instance, had a master's degree
in astronomy, Russell Owen had a master's degree in physics,
Ed Mannery had a bachelor's degree in psychology. But they
were ingenious and highly competent, and they worked respect-
fully together and could disagree without taking umbrage. They
wanted to build things that worked and liked to tell stories
about designs that were expensive and complicated and never
worked. They took great pride in their work, and they wanted
to take long enough to do the job right. They liked things to
be reliable and, after that, simple. Jim had already done much
of the telescope's design, and the UW engineers thought everything Jim put in had a good reason for being there. Listening to
the UW engineers could give you the impression that engineering is the physical incarnation of logic, of sweet reason in the
universe.

On the whole, however, they didn't take to being managed.
Mannery described himself as "a one-motorcycle guy used to getting my way and I don't like bureaucracies with their meetings,
approvals, reviews, and compromising for the sake of compromising." Nor did the UW astronomers supervise them: the UW
astronomy department was small and mostly uninterested in the
survey, and supervising engineers was in no way the astronomers'
job; the UW engineers had simply taken on independent, outside
work that was no one's business but their own. So alone out there
on the Pacific coast, the UW engineers did preliminary designs
for their assignments, sent contracts out to vendors, and what the
vendors didn't construct they sent out to a little blue machine
shop at High Rolls, on the steep road up to Apache Point. They
were under pressure to build it good, build it fast, and build it
cheap, but as always in engineering matters, you can't have all

three, you have to choose two. They chose cheap and good. They didn't consider themselves great schedulers.

In the summer of 1992, the Sloan survey team was composed of four institutions with smart, competent astronomers who knew and liked one another; a national laboratory that knew how to get big projects done and handle huge amounts of data; a clever, competent group of engineers; enough institutional interest to see the project through; reasonable hopes for an NSF grant; and enough donors and the Sloan Foundation to keep money flowing in the meantime. The actors were all on stage, all knew their parts, and were in their places; all they had to do was put on a play. They'd begin by building the hardware. And things, as they do, began going wrong.

The worst set of problems was with the primary mirror. The country's best mirror maker was arguably the University of Arizona's Mirror Lab, but they wanted to build large mirrors in inventive ways, and for them, a 2.5-meter mirror was uninteresting. So Jim, Don, and the UW engineers had contracted the primary mirror out to the Mirror Lab's commercial offshoot, a small company called Hextek. Hextek melted glass, poured it into a 2.5-meter mold, and began the long, slow cool-down to solid glass. But Hextek had no camera with which to monitor the cool-down, and on the morning of July 27, 1992, when technicians opened the oven, they found three cracks in the mirror blank's surface, one at the noon position, one at three o'clock, and one at nine o'clock. The mirror was a sandwich of a faceplate and a backplate with a glass honeycomb between them to make the mirror lighter. The cracks went through the faceplate but not down into the honeycomb.

This had never happened to Hextek before, and at first no one knew what had happened. Mannery guessed—correctly, he found later—that the cooling hadn't been slow enough. The solution seemed obvious: leave the mirror in the mold, put it back in the oven, remelt it just enough for the glass to soften and the cracks to heal, and this time, cool it correctly. If the reheat failed, a new mirror would cost $325,000 and eight months to a year more; happily, it didn't fail.

The newly made mirror blank next needed to have its flat surface ground into a roughly concave shape, a process called curve generating. For that it was sent to a subcontractor who had worked on Apache Point's 3.5-meter mirror and whose shop was set up in his garage in the Arizona desert; he'd done good work early on and was nearby and charged less. The UW engineers, Don, and Jim all had qualms based on the subcontractor's recent work on the 3.5-meter—"a silent scream went up," Jim said—but everything seemed fine until Jim got a phone call that once again, the mirror had cracked. The grinder had lowered itself slowly over the mirror blank, which was rotating under it. But somehow, some plug got pulled and the power to the grinding wheel failed. The grinder stopped moving but didn't stop lowering onto the mirror, and the mirror kept turning against it. The result was a perfectly cylindrical crack through the frontplate, through the honeycomb, and nearly through the backplate.

And again, the fix was simple. In the center of the primary mirror, through the front- and backplates and the honeycomb, is a largeish hole: light hits the primary, is bounced straight up to a smaller mirror called the secondary, which focuses the light back to a camera positioned behind the hole in the primary. The crack was just under three and a half inches from the central hole. So they just ground the central hole that much bigger, losing maybe a couple percent of the mirror's area.

The last step in mirror making is polishing, and for that, the primary was sent to the experts at another University of Arizona enterprise, the Optical Sciences Center. Somehow during the polishing the mirror was supported in such a way that the pressure on its backplate was slightly uneven, and the mirror was polished so it was a highly accurate shape of a potato chip. And of course, it wouldn't come properly to a focus; it was astigmatic. The astigmatism was subtle, around 230 nanometers, or 0.000009 inches. This time, the fix wasn't as simple: it required putting twenty-four small air pistons called actuators around the edge on the back of the mirror and pushing and pulling the mirror in tiny adjustments, maybe 10 pounds of pressure each, until it was shaped correctly. The fix didn't need to be done until the whole telescope was put together, so the Sloanies asked the Optical Sciences Center just to deliver the mirror as is, and the engineers would rig the actuators once the mirror was on the telescope.

The problems with the primary might make its builders look hapless. But mirrors are hard to predict and control. Opticians, the people who make mirrors, are few and don't have much competition, the technology is always being improved, and the procedure is often cut and try. The blanks are glass, and glass is not pure silicon but a composite material, so each blank is different— the glass varying from blank to blank and within each blank—and behaves idiosyncratically. But the finished mirror can tolerate no eccentricity: it must focus all its light in an unwavering point that is 0.25 arcseconds, or 0.0006 inches, across. And to do that, it must conform perfectly to its designed shape, maintain an even temperature throughout, and be polished to a smoothness of 100 nanometers—that is, no roughness higher than 0.0000039 inches. Jim and the UW engineers both love stories about opticians— opticians who have to be bailed out of jail, opticians who are certifiably crazy, opticians who lose an arm either in a silver nitrate explosion or a bar fight, who are often alcoholics. "Apparently the last submillionth of an inch of glass gets into your soul," Jim says, "and only alcohol will wash it away."

In August 1994, the Sloanies all converged for the first time. They'd just had a flurry of successful fund-raising. The Sloan Foundation, the universities, and the National Science Foundation had all promised money. Then a group of Japanese astronomers asked to join the collaboration, offering to pay outright for the CCDs, help out with the software, and to help Jim, send along Maki Sekiguchi, a bright young astronomer who had designed and built a multi-CCD camera in Japan. And the UW astronomers and the U.S. Naval Observatory had also joined.

So the Sloanies, maybe fifty of them now, met for ten days at Yerkes Observatory in Lake Geneva, a small resort town north of Chicago. They worked all day, then came back after dinner and worked until eleven that night. By now, they'd decided to add to the survey a few slices of the southern sky, partly because in the fall the Northern Galactic Cap is not overhead, and partly to use a strategy different from the northern survey's: to image those slices over and over, thereby adding up the images' light, "a gold mine for studying really faint things," said Jim.

At Yerkes, they got to know people whose names they'd only heard, and they began to feel, maybe for the first time, they'd actually created a working collaboration. Jill Knapp got T-shirts made, with a galaxy—not a very good one—on the front and the engineering drawings of the camera, spectrographs, and telescopes on the back. For the occasion, Rich brewed beer, which he labeled Twisted Fibre. The beer's label was specially designed with a medieval-looking astronomer looking at a sky full of Bethlchem stars, and beyond it, a wedge-shaped computer simulation of the universe's large-scale structure. The Sloanies were relaxed and working intensely on the survey.

The survey was progressing more slowly than budgeted for.

They needed to be ready for what they called the test year, a full year in which—once the telescope and mirrors were completed, all the instruments installed on the telescope, everything properly controlled and working together—the data stream would begin and the software to get the data analyzed and sent to the archive would be tested. The Sloanies had originally scheduled the test year to begin in June 1994, then told the funders it would begin the following November. But the primary mirror had had its misadventures, and as of December 1994 was still being polished and would probably take another six months. Tektronix was running late making the CCDs, and Jim wouldn't have the full complement needed for his camera until December 1995. On top of this, the Sloanies now suspected that the UW engineers would need an extra $350,000 for the telescope, plus $270,000 for other items, none of which were in the budget, and the contingency money was already used up. The cost of reaching the end of the test year was estimated at $35 million.

The person in charge of the test year was an Institute postdoc about to be hired at Princeton, Michael Strauss. He began writing a series of test-year updates, which he e-mailed to all the Sloanies, each e-mail moving the test year dates back farther. In November 1994, when the test year was to have started, he wrote, "We are now being pressed for a January 1996 starting date, which, given the summer weather at Apache Point, and the inevitable inefficiencies of commissioning two new telescopes and three new instruments, means that our Test Year is rapidly being reduced to a 'Test Sucker Hole.'" A sucker hole is aviation jargon that was taken over by astronomers. It's a brief opening in an otherwise clouded-over sky that will, just as you fire up your telescope, certainly close.

And meanwhile, nobody was doing much science. Jim's name was appearing on papers, some of which he'd worked on, some he'd only thought about a little, but he'd "basically disappeared from the science scene," he said. People who knew his work but not him personally began wondering what had happened to him—did he die? Other Sloanies in the same situation thought the compensation for scientific invisibility was going to be that the minute the Sloan started producing data on those million galaxies, they'd all write a million papers. But until then they couldn't do science and get the survey done too, so they didn't do science. They were putting on the play like a bunch of neighborhood kids—no hierarchies, no bosses, all working the way they knew how and thought best.

Chapter 4

Photometric War

Jill: Is the software doable, Robert?

Robert: Yes, I think it is. Mind you, this is an unsolved problem in astronomy.

Jim: Everything we've done is an unsolved problem in astronomy.

Back at Princeton, having designed the spectrographs and telescope, Jim went down to the Peyton Hall basement, closed the door, and started work on his camera. Morley Blouke's CCDs, thirty of them, were to be arranged in a grid with six columns, five to a column. In every column, each CCD would be overlaid with a filter that let in only certain colors, so that every image of every object would be recorded in five colors—different kinds of objects are brightest in different colors. But the CCDs were taking a long time to manufacture, and at the moment, the electronics controlling their readout needed to work faster. Jim would eventually put together a small team but for the time being, he was solving problems alone.

One fall day an undergraduate named Connie Rockosi walked into his office and introduced herself. She was nineteen and a junior, stood like a Greek kore, and was nonintrusively friendly. She was a New Jersey local who had grown up building radios and model ships with her father and was surprised when Princeton let her in and gave her financial aid. She was in love with astronomy. When she was still in high school, she'd gone to an amateur astronomy club and seen through a telescope an occultation—a blocking out—of a star by the rings of Saturn. Between Saturn's rings are gaps, which Connie had known about but never seen. What she saw that night was the star disappearing behind a ring, then popping back up, then disappearing behind the next ring— evidence of a gap unseen. It was by far the coolest thing she'd ever witnessed.

The summer of her sophomore year, she interned with an astronomer at Bell Labs who needed someone to help build a telescope's drive system. She had the time of her life: she learned to solder a connection to meet a NASA-qualified technician's approval; she found that the interconnectedness of a circuit was easy to hold in her head; she had a native sense of a network. Now she was an electronics engineering major and needed a subject for her junior independent project. Her Bell Labs boss had suggested she ask Jim for ideas—one of Jim's reputations was for making unique, tricky circuit boards. Could he help her think of a project? Jim talked to Connie for a half hour and thought she was smart and seemed to know what the hell she was doing. He liked that she had been an amateur astronomer and thought that she was, on the whole, a gift from God.

He gave her the job of designing a circuit for his camera, which she did the first semester of that year; then during second semester, she built it and it worked. Jim wasn't surprised; he'd seen that she picked up new things quickly and had a great and

indispensable talent for finding out what's wrong with some finicky electronic bit, figuring out why it went wrong, and fixing it
so it wouldn't ever go wrong again. "She is absolutely dogged,"
Jim said, "and will *find* it and *fix* it." Connie said what she had was
"obdurate patience." A year later, for her senior thesis, she built a
fancier version of the circuit board for Jim's camera. By that time,
she suspected that she was into the Sloan Digital Sky Survey, she
said, "Hook. Line. And sinker."

The camera, the mirror, and the rest of the hardware were all running late but nowhere near as late as the software. Astronomical
software had traditionally been fairly easy to write, but the Sloan
software was de novo: nothing like it had been written before.
The Sloan itself was a new creature, effectively a real-time robotic
astronomer. The software needed to observe and record, then
locate, characterize, identify, and archive. It needed to be able to
handle at least a trillion bytes of data—a terabyte, a unit astronomers had never before had cause to use—coming off the telescope
at a rate of 17 gigabytes, 17 billion bytes, every hour. It needed to
command the automation of the telescope and the observations so
that the only people on the mountain would be a staff of professional observers to oversee and troubleshoot, and everybody else
could stay home and download galaxies. It needed to take the data
coming off the instruments—the camera and spectrographs—
and reduce it, that is, turn data from the CCDs into images and
spectra, standardized so that the stars and galaxies and quasars all
looked as though they'd been taken on the same night under the
same conditions.

The Sloanies were apparently going to have to write most
of their own software: nothing currently available could handle
Sloan's large amount of data with the reliability and autonomy

required. Fermilab routinely handled large amounts of data for huge physics experiments, but that software was too specific to physics experiments, which, for example, sorted through masses of data to find certain bits and discarded the rest. The Sloan software needed to keep everything that hit the detectors.

Software was Fermilab's job. Fermilab had committed five and a half full-time employees to the software and was hoping for help from those free faculty from other institutions Don York had written about. The person in charge of software at Fermilab was Steve Kent, who'd been at Harvard and had been uncertain he wanted to leave academia for a dingy government workplace until he saw Fermilab's glass atrium. The free faculty weren't showing up, so the first thing he did was to supplement Fermilab's five and a half software writers by hiring postdocs, and so did Chicago, Princeton, and Hopkins; the Institute already had salaried postdocs who were free to write software if they wished.

Kent, though an academic, understood that the Sloan software required him to manage an empire, so he made a management plan. In it, different parts of the software, called pipelines, would be parceled out to different institutions. The pipeline that read out the positions of all the objects on the sky, called the astrometric pipeline, was to be done by the U.S. Naval Observatory, which had a tradition in astrometry for navigation. The pipeline for handling the data coming off the spectrographs, the spectroscopic pipeline, was to be done by Chicago, and the pipeline that handled the data from Jim's camera, the photometric pipeline, was Fermilab's.

Then Kent broke each pipeline into a series of steps called modules to be written by people at different institutions. Each module team had a single point of contact, a pipeline coordinator. The module writers were mostly postdocs who were, after all, trained as astronomers, so their software-writing talents were uneven. The pipeline coordinator and his or her team spread over several institutions met regularly, usually by teleconference (which they always called "phonecons"), sometimes face to face. The meetings hadn't been going well. The groups included a lot of procrastinators, many natural leaders, and, as is standard in academia, no bosses, not even Kent, who had no hold over people in other institutions.

Toward the end of 1994, even with the bumpy meetings, Kent's plan was complete enough that he thought he knew what the software would need to do and what resources it would require. He figured that by the time the computers were bought and the software had been written and tested, the survey would be short 73 man-years of people; at $60,000 a person, that came to $4.4 million. Kent had begun by thinking Fermilab had signed on for a heck of a lot and by now was suspecting the Sloanies were all a bunch of rank amateurs.

The coordinator of the pipeline that handled the data from Jim's camera, the photometric pipeline, was a Fermilab postdoc named Heidi Newberg. Heidi was a young physicist who had done her doctoral work at the University of California, Berkeley, on a large project to look for the distant supernovae that were turning out to be another good standard candle by which to measure the universe's expansion. Part of her job on the supernova project had been to write software, so a natural extension of her career was a position as a postdoc at Fermilab writing software for another large project. She knew her career track wasn't going to be fast; she understood that large projects could change the world but that doing them takes time.

The photometric pipeline was the biggest and most complex and most difficult piece of the survey's difficult software. In October 1992, before Fermilab had taken on the job, Jill and Michael Strauss had gone to a blackboard in Princeton and begun sketching it out. It translated a pattern of counts in a CCD's pixels to an image with a certain shape, size, color, and brightness. Jill and Michael's list got long fast. It turned into a yes/no tree that began with finding each CCD's known characteristics and defects and masking out the defects so they didn't confuse the data. Then it got rid of cosmic rays. Then it measured how the atmosphere blurred the objects' images. Then it found the objects and measured their brightness, shape, and position. It measured which objects were transient (meteors, airplanes, asteroids) and which were fixed (stars, galaxies, and quasars). Then it merged with the outputs of other pipelines which translated an object's brightness on the CCD to its real brightness and its position on the CCD to its position on the sky. And finally, it funneled its data into the target selection pipeline, which decided which stars and galaxies should be reobserved with spectrographs to get their spectra. As the blackboard filled up, Jill and Michael wondered how this stuff could be done at all, let alone fast and by postdocs. They typed up their list and sent it to a site for a new means of communication, an archive to which everyone could send e-mails and read the e-mails other people sent.

The person who wrote the software for the e-mail archive was probably the person who should have coordinated the photometric pipeline, an astronomer and software writer at Princeton named Robert Lupton. Robert had been Jim's graduate student at Princeton and then and now, wore no shoes, not even in winter. After finishing his doctorate, he'd wandered briefly into academic astronomy and back out again. His genius was not for astronomy but for writing software, writing code. Sloanies competent to read Robert's code say he's brilliant, an artist, his code sings. "I think I've got good at this," Robert himself says, "so I put myself near the top, probably not at the top. I'm quite good."

But even Robert's friends and admirers describe him as "difficult" and "abrasive," and no one thought he'd make a good pipeline coordinator. Nobody else seemed to be offering to do the job though, so Heidi Newberg volunteered to be the photometric pipeline coordinator and sent out e-mail to that effect. The people on her team included postdocs at Hopkins, the Institute, and Princeton, some scientists in Tokyo, and Robert.

Heidi and a computer scientist at Fermilab worked up a version of the computer language Tool Command Language, or Tcl, that was called Fermilab Tcl or F-Tcl, and pronounced "f-tickle." She assigned modules to various people and they got to work on software that would, variously, find objects in an image, or separate objects light-years apart that appeared to lie on top of one another, or measure how spread out the object's light was. Heidi was trying to put together not a full working pipeline but what was called a level 0—a design, a backbone, an infrastructure—in which sketchily written modules would be placeholders for the real things. Then the module writers could iterate, to go on to level 1 and level 2, improving the pipeline each time through, adding features where needed, speeding it up.

That was Heidi's plan. She and the photometric pipeline team communicated with phonecons and on the e-mail lists, Heidi asking brief questions and provoking long discussions and disagreements, often involving Robert. Heidi would give Robert an assignment, Robert wouldn't do it. Robert was impatient and English, so he expressed impatience orally in an extremely precise accent and in e-mail with lethal formality. Once Heidi e-mailed out a question whose answer had already been covered in an earlier document. Robert e-mailed back publicly, "I hate to sound as if I come from Fermilab, but the reason is given in the Functional Specification for Data Processing . . . I believe that the point that you should Read The Manual (abbreviated RTFM in unix) is worth making." Heidi wrote back another public e-mail titled "Respect" that was a stiff lesson on manners and added that one of her team "expressed angst about whether his code would be sneered at by the rest of the collaboration."

The disagreements just went on from there. Ordinary discussions of every ramification and aspect of some technical question always seemed to veer off into an underlying disagreement, usually about quality, and usually Fermilab was on one side and Princeton was on the other. Fermilab thought the job should follow a plan and be done well enough to meet the deadline. Princeton thought it should be done right, no qualifiers. Fermilab, knowing that spectroscopy could be done only on bright galaxies, seems to have suggested at one meeting that the photo pipeline not capture every bit of signal out of the noise in order to find dim galaxies. Princeton was horrified. Jim was not on the photo pipeline team, but the pipeline was the medium through which his camera spoke to the world. The camera could detect objects thirty times fainter than the proposed pipeline would find. He was going to exquisite lengths to build an instrument that took extraordinary amounts of data with extraordinary precision and sensitivity, and he wanted no galaxy lost.

And how should the software treat numbers? Should you write 2.3, for example, as an integer, 23, and remember to divide by ten, or should you write it with a floating point as 2.3? Heidi proposed using floating points: they were slower and took twice the memory but wouldn't require what she called "cleverly crafted code." Princeton felt that speed was essential and said that floats required more memory than Fermilab's computers had. Neither

Kent nor Heidi disputed any of this, but they thought that by the time the survey got going, new computers would have the speed and memory required, and besides, computers were cheaper than software writers.

Jim—again, not on the pipeline team but worrying about his camera—felt so strongly about floating points that he wrote one of his e-mail screeds: "The photometric pipeline is, I think, the single most difficult computational job ever attempted in observational astronomy, and many much simpler jobs like it have failed." And then he became personal: "The survey project is certainly no longer 'mine'—it has grown too much and changed too much, but I still regard it as my child; the data processing is crucial to its success, I do not want it to fail because of a very foolish choice like this, and I think it is likely to." Fermilab's Kent thought that Jim occasionally lost a sense of proportion about what was important and what just wasn't worth worrying about. Princeton's Robert felt they were facing death by a thousand cuts.

Heidi didn't like Princeton's assurance that it knew the right way to do things. In the Fermilab culture, she thought, compromise and trust were necessary, and sometimes you went along with what someone else did even if you thought you could do it better. Besides, she thought, Jim seemed to want no limits on the pipeline at all; he wanted it to do everything possible. In the meantime a Princeton postdoc on the photo team named Michael Richmond had been regularly running tests of the photo pipeline on the Fermilab computers and found that the pipeline wouldn't be able to keep up with the data coming off the camera, that it ran too slowly by a factor of five. In July 1994, Kent reported that the level 0's of all the pipelines should have been finished seven months before, that deadline had now slipped, and no new one was being mentioned. Shortly afterward, Kent asked Heidi if she'd be OK with the photo pipeline being taken away from her and turned over to

Princeton. She said she wouldn't be upset, but she did wish the level 0 photometric pipeline could be considered, at least, a good first try.

So in August 1994, Jill became the photometric pipeline coordinator, and though she was no better or worse than any other astronomer at writing software, she thought Robert could do the pipeline pretty much by himself: "Robert is a difficult son of a bitch actually," she said, "but unbelievably, oh my goodness, talented." Jill thought Heidi had been in a miserable situation with a job that was de facto impossible. Robert would write the pipeline with the help of the Princeton postdoc Michael Richmond, and Jill would manage them both. As a management strategy, Jill scheduled weekly pipeline meetings at noon and provided a countertop full of food. Princeton and Institute postdocs started showing up for food and stayed to work on the photo pipeline. Eventually the photo pipeline was replaced entirely; the final version did much more using a tenth of the memory and time.

Kent decided that keeping the pipeline teams and their coordinators in the same institution just worked better. That way, he thought, the team was accountable to the coordinator and when a pipeline wasn't working, everyone knew whom to blame.

Sloanies had clearly and admittedly underestimated the software—they all say so. Even if they'd known what they were up against, software is notoriously difficult to manage, to work out the time and number of people needed to produce code of a certain quality. The fundamental difficulty was distinguishing what is good from what is good enough. Software writers come in an enormous range of talents—Robert liked to quote a claim that the top software writers are a factor of one hundred better than those at the bottom, then added, "I would say it's minus one hundred myself because some people make a small negative contribution by doing things badly." Kent thought that one very good software writer accomplishes more than ten good-enough writers. But how to find that one very good writer without hiring ten writers and winnowing them down? Kent once asked a professional science project manager how he managed software, and the professional said, "I don't do that." Software management was and still is a problem with no good solution.

Adding to the tension over software was the problem with the Sloan postdocs. Postdocs in general have a few years, usually three, to do splendid science that will get them splendid jobs. Sloan postdocs were, in principle, paid to spend half their time on survey software and the other half doing their own research. But in practice, many of them spent all their time writing code. The point for the young postdocs was that they wanted to do science, they knew how to write software, and their payoff would be terabytes of data on every object in every part of the heavens. When they were hired by the survey in 1992 or 1993, they were told that data would start streaming in late 1994.

That clearly wasn't happening. Heidi had gone up to Apache Point, and the telescope was just a hole in the ground the size of a hot tub, and a telescope engineer with back problems was lying on the grass next to the hole. Heidi thought probably her career was over and she should go get an MBA instead.

"Things just aren't getting done," said Jim Annis, a postdoc at Fermilab. "There isn't a telescope, there isn't going to be a telescope, there's just nothing. Nothing, nothing, nothing occurring."

"We were told every year, beginning in ninety-three, we'll have data in a year," said Bob Nichol, a postdoc at Chicago. "Then they said next year, next year, next year."

Jim Annis said that every time three Sloan postdocs got together, a Sloan support group formed: "Everyone else goes to a conference to talk about science. We go there and just talk about the Sloan. And what we could do. Someday."

Without data, how could they do science? With no science, how could they get jobs? In late 1994, Tom Nash of Fermilab, Jill Knapp at Princeton, Tim Heckman at Hopkins, and Mike Turner at Chicago formed a committee to report to ARC, the survey's governing body, on youth and the software crisis. The Nash Committee, as it became known, saw the problem as only partly about the postdocs; the rest of the problem was with the senior astronomers, the professors who had institutional rights to the data whether they worked on the survey or not. Early in the survey, astronomical faculty at all the universities had signed up to be on working groups—on galaxies, or clusters, or stars, or quasars—which were to help decide the survey's science. And since that science would get to these professors via the software, Don York and the rest of the survey management had hoped the professors would volunteer their time and help write the software. After all, Jim, Rich, Don, etc. weren't being paid a penny either, and certainly software needed to be written.

But the working-group professors not only were not writing software, they weren't attending software meetings, and in too many cases they weren't working on the working groups at all. They divided in general between a few people who, like Jill, got 100 percent into it and many more people who went about their normal lives until the data came in. For these professors, the decision was the same as the one about the oversight of engineers: no one's job included writing software for the survey, nor was anyone being paid to do it, and writing software was no more important to an astronomer's résumé and reputation than was building instruments. The Nash Committee floated some solutions to the professorial nonparticipation problem, but senior professors operate as independent and locally owned subsidiaries, and no solution seemed workable.

So the Nash Committee concentrated on the postdocs. They held two open meetings to collect grievances then wrote it all up in a balanced report. They listed the survey jobs needing the attention of the professors, urged the professors to talk to the postdocs, and recommended that postdocs leaving the survey for a job be able to carry their rights to the data with them. The report was long. They sent it to Jerry Ostriker, who in addition to everything else was also on the ARC board and had commissioned the report in the first place. Jerry and the Nash Committee met with the full ARC board, Nash explained the report, they all argued. Hopkins's Tim Heckman sat in the meeting thinking that this had been a waste of time. On the whole it had been; after the meeting, he said, "it was over and done with."

A little later, the ARC board instituted a policy about the portability of data rights that reassured the postdocs. Otherwise, nothing changed. More than a year later, Jim Annis wrote a group e-mail: "If I may ask, how do you expect us to get astronomy jobs if we: a) concentrate on making the infrastructure, and b) publish only a few papers and those in alphabetical order? Will the kudos of the collaboration suffice?"

Chapter 5

Running Open Loop

In a five-year [astronomer]-led project, at year five, month eleven, you go, "Oh! Shit!"

—Michael Turner, University of Chicago

You have inspiration guys and science guys and charisma guys and community builders. But the guy with the work breakdown structure, he's the grown-up.

—Craig Hogan, University of Washington

THE SLOANIES HAD been giving talks to astronomy departments and conferences, claiming that the survey would revolutionize astronomy. In return, the astronomical community was being snide, doubting that the Sloan would ever take data or, if it did, would ever share it. The community worried that the considerable amount of money the NSF was putting into the Sloan would mean less money for their own projects, and they may have been right. They worried that they'd write proposals to use telescopes and the observatories would write back that Sloan was going to do their proposals better. They may have been right about that too, but they thought doing something better is always easier when you haven't done anything yet. Sloan was making enemies. Sloan was becoming a dirty word.

Then in early 1995, the Sloanies began hearing about a project by English and Australian astronomers called the 2dF, for Two-degree Field, a wide-field survey that would find the redshifts on 250,000 galaxies. In March, the Princeton postdoc Michael Richmond ran across an Australian astronomy magazine with an article titled "Mapping the Universe" that promised the 2dF survey would begin shortly and announced, "The race is on." Bob Nichol, the Chicago postdoc, told the Sloanies he'd seen an official paper from 2dF, and the third sentence promised it would beat the Sloan in just two years.

A Hopkins Sloanie, Alex Szalay, was a personal friend of some of the 2dF people and was upset they'd been so secretive and were now declaring a contest. He suggested that the minute everything was installed at Apache Point, the Sloanies do a small redshift survey, the SRS, and find the large scale structure even before 2dF did. The Sloanies discussed Alex's SRS to pieces but finally decided to stop panicking and to concentrate their efforts on getting the Sloan on the sky. Jim Gunn just shrugged and said he was "positively disposed toward anything that would give us a kick in the rear."

The 2dF survey went on to do as it promised, to make a map big enough to find the size of the large-scale structure. For the first time, the edges of the map—2 billion light-years away—were bigger than the largest structures, which were 300 million light-years across. The hierarchy of galaxies, clusters, and superclusters went no higher. The map of large-scale structure looked like someone had scattered iron filings over a plate then briefly run a magnet underneath: no particular pattern, but in no way patternless.

2dF had limitations. It was going to cover about 1,500 square degrees of sky versus Sloan's 10,000 square degrees; 5 percent of the sky versus 25 percent; and get redshifts of 250,000 galaxies versus Sloan's million. It took no images but did spectra only, and only of galaxies selected from photographic plates, a strategy that Jim had specifically rejected in 1987 as less accurate and less systematic. John Peoples, Fermilab's physicist-director, listened to the Sloanies compare 2dF to their own survey—extraordinary numbers, completeness, imaging-plus-spectra, selection effects, and systematic errors—and said, "That's when I understood what Jim wanted to do. A really good survey."

2dF made clear what the Sloanies knew all along: that despite their own stated mission of finding the large-scale structure and as compelling as that mission was, their survey was turning out to be more various and revealing. They were going to be able to do what Jim's career had been aimed at: to find, name, and understand all inhabitants of the nearby universe.

Future glory aside, the inglorious present still included endemic money problems, chronic lateness, disillusioned postdocs, tribal warfare, scientists not doing science, and a test year sucker hole. The NSF had finally agreed to send money, but only if the Sloan addressed these problems by hiring a professional project manager. Most Sloanies weren't sure what a project manager was; management wasn't something astronomy projects needed, the partners just went off and did what they were good at. To the extent that Sloan had a management, Jim was the project scientist and Don York was effectively both project director and project manager.

So under duress, the Sloanies hired a professional manager named Tom Dombeck. Don York gave him no authority and no money. Dombeck began by putting together a budget and what project managers call a work breakdown structure: what the jobs were, what the steps in the jobs were, who does the jobs, how long the jobs take, and what the jobs need someone else to have done first. Dombeck wrote it up, everybody agreed to it, they presented it to the NSF, then, Dombeck thought, Don put it in a file cabinet. Dombeck set up regular meetings of a group called the Installation Task Force in charge of getting the telescope and instruments installed on the mountain in time to begin the test year, which by then was two months in the past. The task force had regular meetings, each meeting detailing the impossibility of getting one piece of hardware after another done on time. After two years of this, in June 1996, Dombeck decided he wasn't earning his keep, couldn't even sign a requisition, and quit.

The same June, a meeting of the photometric pipeline group at Princeton began with a thought for the day: There Is Too Much to Do. By September 1996, both the primary and secondary mirrors had been delivered to Apache Point, but because the telescope was not yet ready for them, they were sitting in crates. The telescope wasn't ready because the UW engineers were swamped, weren't getting things done, and ceded hardware jobs to Fermilab. In October 1996, one of the Japanese astronomers wrote to the email list, "This is a completely unscheduled project, just like Italian trains; we were already 2½ yr behind and no one knows the real schedule." In November 1996, Jim said that the Fermilab group now in charge of the telescope's controls might be more or less done by late January 1997.

Fermilab, though running late, was making progress with the hardware for the telescope controls, but the software was being done by someone who didn't necessarily understand what was required and Fermilab didn't want help. Nor did Fermilab improve matters by sending to Apache Point a physicist named Steve Bracker—described variously as brilliant, a cowboy with a

thousand-mile ego, highly competent, a thorn in everyone's side—
to work on the telescope controls. The UW engineers felt Bracker
was also there to keep an eye on them—that he was highly critical
of them. Ed Mannery said Bracker wouldn't care if Jesus Christ and
all the apostles said something was OK, Bracker wouldn't go along
with it unless he thought so too. But given Bracker's immense
skills, Don York thought Bracker would be hard to replace.

Rich Kron had decided earlier to spend the summer of 1996
at Apache Point. He thought the timing would be good because
they'd have installed the instruments on the telescope by then. So
he made the necessary arrangements, rented a house near Apache
Point, but when he got there, the telescope had no instruments, no
mirrors, and couldn't be controlled. So he spent a quiet summer
in the New Mexico mountains, working on unrelated things. In
fact, though Rich was still trying to be generally helpful, he was
pulling back from the Sloan a little. His title—to the extent that
anyone had meaningful titles—was survey director, and he wasn't
sure what that job was; he didn't think it was really a job, and he
was also personally allergic to squabbles, and the Sloan these days
wasn't so much fun.

The project manager who replaced Dombeck was an astronomer from the U.S. Naval Observatory named Jeff Pier. He had
no more authority or budget than Dombeck had and no management experience whatever. He watched while every day the schedule slipped by a day, every month by a month. The NSF saw the
same thing, and since it had been parceling out its award year by
year, decided to withhold the last $1 million. The NSF's program
officer could be paraphrased as saying that he wouldn't give the
Sloanies any more money until he was convinced it would be well-
spent and "sunk cost" was not a good argument.

In Pier's weekly flights to Apache Point over Phoenix, he
watched a ballpark being built—cement trucks busying around,

girders going up—and saw nut-and-bolt details gradually taking shape into a giant stadium and thought, "And we can't even build a two-point-five-meter telescope." He was losing sleep and getting frightened. He thought they were in such a mess he didn't even know where to begin. Here was an important project which he believed in very, very much, and here he was, unable to handle the job; it was just more than he knew how to do.

The person who knew what Pier didn't was Fermilab's director, John Peoples. Peoples had trained in the arrogant, competitive, aggressive world of high-energy physics and had been part of its enormous cultural shift: in the early 1960s, for his thesis experiment, he worked with 5 people; his follow-up experiment had 20; by the time the series of experiments was complete, in the late 1990s, he was working in a group of 120. You don't go into experimental high-energy physics, he said, unless you're prepared to work with other people. He'd worked in the aerospace industry, then made his way to a directorship in a national laboratory hierarchy in Washington, D.C., that required directors to face congressional snipers and the wolf pack that was the directors of the competing national labs.

Peoples has amused and calculating eyes, a soft voice, thinks in strategies, and knows where the power lies; he is a thoroughly political human, though unlike others of the kind, he does not talk circuitously to soften his intent. He'd just come back from closing down the Superconducting Super Collider, particle physics' biggest experiment, after Congress pulled its funding. In nine months, he'd reduced the work force from 2,400 to 400—and he hadn't found this difficult to do. Now he needed to start paying attention to the Sloan. Fermilab had made a multimillion dollar investment; he didn't intend for the lab to fail at its first venture outside of experimental physics. He thought, "Jesus, there must be some way to save this thing."

In late 1995 he went out to Apache Point and as he said, kicked around at the tires. He learned that the hardware and software were both interminably late. He went back and talked to Fermilab people who told him that the head UW engineer was just telling Don York whatever Don wanted to hear and the project was a disaster. He sat down and wrote a letter to Don pulling Fermilab out of the project, then never sent it.

By now, Peoples, whose official position in the survey was chair of ARC's Advisory Council that oversaw the survey and director of one of its member institutions, decided he wanted the lateness quantified. He couldn't believe it hadn't been done before, but he didn't know about Dombeck's work breakdown structure in the file cabinet. He asked Fermilab's Steve Kent, Mike Evans at UW, who was ARC's business manager, and a Fermilab engineer/manager named Bill Boroski to go around the country, talk to everybody, and make a list of what had to be done to get the telescope on the sky.

Princeton called Kent, Evans, and Boroski the tiger team. They went to Apache Point, UW, Princeton, and Hopkins, sometimes repeatedly; Boroski said that over the next two months, he probably slept at home one week. He found the Sloanies open and helpful, dedicated to the project and saying, in effect, "We have a lot of work to do but we know we can do it." At each stop, the tiger team asked all the questions that go into a work breakdown structure. At a time when the Sloanies were talking about the two-years-late test year beginning in a few months, the tiger team found the actual amount of work left would take another two years.

Jerry Ostriker was hearing talk about people bailing out and thought, "When I have a viable project, I'm always sure I can raise the funds, but at this point, I'm not sure we have a viable project. I think the thing might just die." He had seen astronomical projects dissolve when the people working on them didn't get paid and just stopped working, and this is how he thought the Sloan might die.

Jim Gunn was glad the tiger team report had been done. He thought it made the necessary case that the Sloanies either had to give the thing up as hopeless or get a real project manager. He had a phrase he liked using, "running open loop." "Open loop" is an engineering term meaning a system that runs without feedback, without a self-governor, without correcting itself. A closed loop sprinkler system has a groundwater sensor to tell it when to stop watering the lawn; an open loop sprinkler keeps sprinkling while the lawn washes down the hill. Jim thought the whole project was running open loop.

His part in all this had been subtle. He may have been the only Sloanie to understand the survey end to end; he carried the whole project around in his head. Not only did he understand the whole project, he probably could have done every part of it—hardware, software, science—by himself. So he felt personally responsible for everything. "Other people are in charge of things," he said, "but I don't really feel like I'm doing my job unless I'm really watching things pretty carefully." "Watching things" sounds passive; Jim was active, he was into everything. He was the survey's central character, its core, and occasionally its bottleneck. Whole programs would come to a stop because somebody needed to ask Jim a question and Jim was too busy to answer. He was doing an unprecedented number of other people's jobs: he liked things done right, and when other people didn't do them right, Jim fixed what they'd done or gave the job to someone he'd learned to trust. He knew the survey so well he was very nearly winging it, solving problems as they came up because he knew the solutions and knew how everything needed to intersect. "But I didn't understand how crucial management was, just in terms of money and time," he said later. "I just didn't understand. I could have understood if I had taken the time to try to understand, but I didn't."

His one official job, besides project scientist, was to build the camera, and the camera was late. The camera was the survey's sine qua non; a project manager would say it was on the critical path. He didn't know whether he could build it faster or not—he didn't think it would take as long as it seemed to be taking. All he knew was roughly what it would cost and that he could do the job and that when he was done, the camera would be as good as a camera could possibly be. He'd never worked to a schedule in his life, he didn't know a work breakdown structure when he saw one. He was—to use another project management phrase—a single-point failure: if Jim got hit by a truck, everyone said, the survey would stop dead. "I sort of look at myself and my own role in this with some horror," he said.

A line from an ARC budget document, dated January 1997, freely translated: Cash needed, including $1.6 million in unpaid invoices, by June 1997: $3,360,000. The "$1.6 million in unpaid invoices" meant that Princeton, UW, Chicago, and Hopkins were paying engineers and postdocs to work on the survey but ARC, the survey's parent, wasn't reimbursing the universities. The universities began pressuring their representatives on the ARC board, which approved ARC's Advisory Council to appoint a special committee, which in turn decided that the tiger team's fact-finding trip should be a precursor to a full-scale review by three outside, neutral people.

The three reviewers asked everyone on the survey's management chart to meet at Fermilab on June 8 and 9, 1997. And they asked everyone else to send e-mails or write letters or make phone calls. At Fermilab, the reviewers met in a conference room called the Comitium—the name of the central place of assembly in ancient Rome—around an elliptical table. They asked survey members to come in, one by one.

What they heard was dismaying but—given what Tom Dombeck, Steve Kent, Jeff Pier, the Nash Committee, and the Fermilab tiger team had already found—not surprising. No one, from Don on down, could estimate credibly how much the survey would cost and how long it would take to start scanning the sky: estimates ranged from one year to three and a half. No one knew whether the survey would ever take data or if it finally did, whether they'd still be around to claim data rights. People were not given specific goals, just told to "do the best you can." Requirements for the software were moving targets: different science working groups would make up new requirements and give them directly to the software writers. Requirements for the hardware were contradictory: asked to what precision the telescope needed to point and track, Don and the UW engineers offered two different numbers.

With minor differences and a few exceptions, everyone seemed to tell the same story: their work was second-guessed, criticized, and done over; if they tried to come up with accurate forecasts of time and money, Don told them not to; they were frustrated, they weren't having fun, they were uniformly gloomy, they'd lost their spark. One reviewer, Jim Crocker, listening to the Sloanies, was reminded of the outnumbered English army before the battle of Agincourt: they thought they'd get massacred, they thought they were all going to die.

A couple of weeks later, the reviewers reported back. Crocker thought the Sloan's situation was predictable. "They want to achieve perfection," he said. "This is their magnum opus. They are trying to achieve something that is exponentially approaching the unaffordable." In previous projects he managed, he'd had to tell scientists they were going to achieve all the perfection they could with the money and time that other people were willing to give them; Crocker had said this so often he abbreviated it as Other People's Money, or OPM. And the Sloan was now going to cost in excess of $70 million in OPM, comparable to the biggest new telescopes, like the 8-meter Gemini or the 10-meter Keck.

With certain of life's crises, the problem is easy to understand but the solution is complicated. With the Sloan's near death, the problem had taken nearly ten years to understand, but the solution was simple: Hire a hard-eyed project manager who would enforce budgets and deadlines. Specify every single science requirement for the survey. Write a full-scale work breakdown structure. Find another project director.

Finding the next director was sure to provoke interinstitutional mud wrestling, and the Sloanies put it off for a few months. But they badly needed to reassure the Sloan Foundation, the NSF, and the jittery universities that they could pull themselves together. So the first thing the Sloanies did, after advertising nationally, was to ask Jim Crocker to be project manager. Crocker was then managing the European Very Large Telescope, but before that he had been across the street from Hopkins at the Space Telescope Science Institute, where he managed the fix that corrected the mirror and turned the Hubble Space Telescope from an easy joke into a publicist's dream. His management record was impeccable, he was impartial and eminently reasonable, he had a homey-sounding Deep South accent, he talked to people as though he'd known and respected them forever, and he had the near-magic ability to solidify a group. Jim Gunn thought he was a breath of fresh air.

The first thing Crocker did after accepting the job was ask Princeton, Hopkins, UW, Fermilab, and Apache Point for their schedules so he could outline the critical path. "Let's say you're building a house," he explained. "I start digging the foundation, I got the crews here. But where's your building permit? Well, you don't know. So everybody stands there while you go get the permit and that might take six weeks. Critical path on a program are those items that you have to start before you can do the next one." He took the schedules, the critical path, the Kent/Boroski/Evans work breakdown structure, and integrated them all on one page.

A survey that had seemed rusted shut started moving. A newly hired UW engineer named French Leger ran into a problem with a large rotating disk, attached to the telescope, to which the camera and spectrographs are attached. The rotator needed to turn extremely precisely and extremely slowly, about the width of two moons every eight hours; instead, it would turn a little, seize up, then jump, as French said, "go whiiiisht"—a classic stick-slip. At a meeting at Apache Point, French presented the problem, Crocker asked for the solution. French said, "We need a new rotator." Crocker said, "Do you have a design for one?" French, who'd just done a design, said yes. Crocker asked how much money and how long. French said he'd need a draftsman, $80,000 to $85,000, and three months. Crocker told him to do it. French was astounded that Crocker knew where to find $80,000 and how to get things done. The rotator was eventually delivered a week ahead of time and about $2,000 under budget.

Crocker assigned Princeton's Michael Strauss, who had been working on the test year since 1993, to write down the survey's science requirements: for each piece of hardware and software, what's required, how you test it, what bad thing will happen if you don't meet the requirements. "Measure the properties of the universe" is not a good science requirement. "Measure the brightness of bright stars to a repeatable accuracy of 0.02 magnitudes or 2 percent" is a good science requirement. Strauss put together a reluctant team—everyone was busy, no one liked writing picky specifications—and Crocker nagged them: a requirement you can't verify is not a requirement, he'd say. He'd ask, "To what level does this component have to perform, to within what tolerance?" They'd say, "As good as it can be." He'd say, "How good is that? As good as the laws of physics will allow and engineers can achieve?" "Pretty much," they'd say. "So write it down for me," he'd say. "Is it .000000000001? or .00000000000000001?"

The foremost proponent of "as good as the laws of physics allow" was Jim Gunn. He pointed out in meetings and in private that writing down requirements for systems that were all but built was completely silly. He pointed this out in e-mails: "Airy-fairy 'requirements' about what may happen five years after a test year which has not happened yet seem to me premature to the point of foolishness." He pointed this out about software: "You just go and figure out how much information is in the data and you extract all of it and you don't need any requirements. If you had a set of requirements, it might tell you to extract more, which would be impossible. Or it might tell you to extract less, which would really piss you off."

He pointed it out about hardware too. The CCDs were the best available, he said, and so were the fibers in the spectrograph, and the 2.5-meter telescope was the size that fit them both. That fit, he said, wasn't rocket science: "You sat down," he said, "and you figured out according to the rules that had existed for a very long time how to match things together and build the best thing you could build from those components." He thought the whole system was simple enough that well-intentioned, intelligent people could sit down and design this thing without some two-thousand-page book to tell them what it needed to do. Remember, he said, he had come to the Sloan from the Hubble Space Telescope, "which was drowning in this shit. I just don't like the idea."

One of the biggest requirements arguments was over an exact and usually uninteresting subfield of astronomy called astrometry, the study of how to measure precisely a star's position in the sky. One of astrometry's problems is that, just as ripples in a stream change the apparent position of rocks underneath, ripples in the atmosphere change the star's apparent position, making the star twinkle. The Sloanies' most recent proposal had specified that stars' positions had to be measured to within 260 milli-arcseconds —any larger error, and one spectroscopic fiber couldn't be placed accurately on one star. But how to set the actual requirements? If the atmosphere allowed positions to be measured only to a certain accuracy, why set requirements to a lower number that the atmosphere wasn't going to let them meet anyway? Jim's idea was, the better the astrometry, the better the science; maybe they could measure the motions of stars over time. Jim wrote, "We should be at the limit beyond which it would be very hard to go." Finally Crocker effectively locked the principals in a room in Fermilab until they agreed. After a couple days, they did: they voted and more or less agreed on all the numbers except the one about the accuracy demanded by the science. Everyone said that stars' positions should be measured to within 100 milli-arcseconds, except Jim, who said 60 milli-arcseconds. They agreed on 100 milli-arcseconds, with 60 milli-arcseconds as "an enhanced goal."

Jim eventually signed off on Strauss's forty-two-page science-requirements document, but he remained unconverted. "It's always simpler and cheaper to do it right to begin with," he said. "Because if you don't do it right to begin with, you'll have to do it again, no matter what the bloody cost and schedule says."

The biggest hardware problem that Crocker tackled was a companion to the 2.5-meter telescope, a small, standard telescope with a 24-inch mirror, called the monitor telescope. Before the survey could start, the monitor telescope had to do its own survey of stars to standardize their brightness—that is, make sure they weren't variables or irregulars, then compare each star to some standard and calibrate the star's brightness accordingly. The monitor telescope would then tell the 2.5-meter how to calibrate the survey stars. This systematic and precise measurement of brightness was to be one of the survey's central virtues.

Back in 1992, the UW engineers had taken two bids for a monitor telescope, one from an established company with a reputation for being hard to deal with and a second from an accommodating start-up called AutoScope whose bid was far lower—at $192,000, about half. In 1995, AutoScope delivered the monitor telescope and the engineers and astronomers at Apache Point started testing it. The AutoScope telescope could do almost nothing right. It had a cover over its mirror that opened with a little motor, but any breeze would blow the cover shut again, and Apache Point sits on the top of a windy cliff. The fork on which the telescope was mounted was aluminum, and when the telescope moved, the fork buckled; if the telescope was told to go to Polaris, and if the east side of the fork buckled, the telescope slid east to somewhere in the neighborhood of Polaris. The motors weren't strong enough to point the telescope rigidly; if the telescope was pointed at Polaris and someone touched the telescope or if the wind blew, the telescope moved off Polaris.

For the next three years, a small team of Sloanies—two postdocs and a technician—dedicated themselves to patching up the monitor telescope and its software. They tried holding the mirror cover open with magnets. They priced a new fork mount. They replaced the entire control system. Crocker figured that all together, they'd spent "numerous man-years" and over $500,000 on a $192,000 telescope.

Delivery of a new telescope from another company would take fourteen months. Crocker put that fourteen months together with the year the monitor telescope would need to calibrate the sky before the survey could even begin and decided that a new telescope and two-year delay would sink the survey for sure. While considering the next option, he happened to be at the Space Telescope Science Institute, which looks over at the physics and astronomy building at Hopkins, Bloomberg hall. He noticed that on top of Bloomberg was a telescope dome and wondered what kind of telescope was in it, so he asked Alan Uomoto, who was building the spectrographs, about it. Uomoto told him it had been a gift from Michael Bloomberg, a Hopkins trustee, and was used to educate the public but not to do science because Baltimore's night sky, being a rich carnelian-colored mix of humidity, haze, and streetlights, never held much astronomy besides the Orion nebula and a couple of planets. Uomoto took Crocker up to look at the Hopkins telescope and Crocker thought that compared to the monitor telescope, which looked to him like somebody's toy, this one was really nice, an astronomical-quality telescope. Crocker then went to see the chairman of the physics and astronomy department, who told him that if the Sloan replaced the Hopkins telescope in a timely manner and if Michael Bloomberg didn't mind, the Sloan could have the telescope right then. Bloomberg didn't mind; in fact, Crocker thought he seemed actually happy to help save the Sloan Digital Sky Survey. Moving the Hopkins telescope and replacing it with another one, Crocker estimated, would cost $298,000.

Uomoto called the Hopkins telescope's manufacturer—which happened to have been the manufacturer whose higher bid for the Apache Point monitor telescope was rejected—and asked them to come move it. They brought in a crane, picked up the telescope, put it in a U-Haul, and Smee and another Hopkins engineer drove it out to Apache Point. "It wasn't rocket science," said Crocker.

The telescope got to Apache Point four days after it left Baltimore, was installed, and worked like a turnkey. It was renamed the photometric telescope, or PT. The AutoScope telescope was packed up and sold for $35,000 to a rich Texan amateur astronomer who wasn't as worried about precision. Meanwhile, AutoScope seems to have run into trouble with optics suppliers and apparently went out of business, and the Sloanies never heard from the company's owner again.

In the fall of 1997, while the Apache Point staff was still writing e-mails titled "MT troubles," the Sloanies started the business of finding a new director. ARC's Advisory Council set up a special committee that screened the candidates, one of whom was Don York, and by mid-November 1997 narrowed the field to two, neither one of whom was Don. Don resigned, which was hard for everyone concerned: he's approachable, good at mediating, patient, and had his whole heart in the survey. Jerry, who had known Don at Princeton and backed him to be director, felt that Don's lack of ruthlessness, not firing people who weren't getting work done, was only because Don was a decent human being. Don himself said he valued collegiality, people's willingness to look out for one another and act for the greater good. The Sloan would be the glory of astronomy because the people in the Sloan project were passionate and perfectionist, Jim especially, and they wanted to do it right. "No project works that way," Don said. "But I tried to make it work that way." He was deeply upset: "I have cotton over my eyes," he said. "I don't talk to anyone about it." To his everlasting credit, he never left the collaboration.

The two remaining candidates for director were Chicago's Michael Turner and UW's Bruce Margon. Turner wanted the survey to follow the physics model of an experiment and concentrate on cosmology—the universe's density, expansion, and large-scale structure. Margon said the survey would find everything from protoplanets to quasars, so it should concentrate on getting everything into the archive. Chicago supported Turner. Princeton supported Margon. The ARC Advisory Council's special committee came down equally on both sides and hit an impasse. The special committee referred the impasse back to the Advisory Council, which hit its own impasse. Word got out to the troops, and what had been an impasse grew to become a severe strain on the collaboration. Finally management agreed on a four-headed compromise: Margon became scientific director, in charge of making sure scientific requirements were met; Turner became project spokesperson, in charge of talking to the funding sources, releasing information to the public, and protecting the Sloanies' morale; Jim Gunn remained project scientist, in charge of everything he was in charge of; and Hopkins's Tim Heckman became temporary project director. Tim, along with a lot of other people, thought this management scheme was goofy and doomed to fail, but it would probably keep things together until John Peoples took over as director, which it did.

Project manager Jim Crocker announced that in 1999 he was going to take a job at Ball Aerospace, where he'd manage the design team for the next space telescope. He didn't feel bad about leaving. He thought the project was pretty much on the right track. "They'd been only half a bubble off," he said. "You know, these carpenter levels? The problem is, when you're half a bubble off and your wall is 6 miles long, at the other end you're off by a quarter mile. So all they needed to do was be moved back by about half a bubble and then everything was kind of fine again."

The Sloanies were sorry to see him go and for years afterward, talked about him. "He came in and just bang bang bang bang, straightened things out," said Jim Gunn. "I mean, he saved our asses. He certainly did things I wouldn't have had sense enough to put together, but he didn't do anything I disagreed with. He was wonderful. He was just bloody wonderful."

The survey, galvanized, had come together. Out at Apache Point, the mirrors had been installed, and then the camera, and the camera had been opened up. It couldn't start surveying yet; it just tested itself and for the first time recorded starlight. Astronomers' lovely name for this moment is first light.

Chapter 6

First Light

First light was fairly absurd altogether.

—Connie Rockosi, University of Chicago

W HILE THE SLOAN was sorting out how to manage itself, Jim was back in Peyton Hall's basement, finishing up his camera. He worked with a small cohort that included Connie Rockosi; the technically inclined Japanese astronomer Maki Sekiguchi; and an engineer named Michael Carr. Michael and Jim had worked together off and on since the mid-1970s when they were at Caltech, building WFPC, PFUEI, and Four-Shooter. Before that, Michael had been a drag racer at the Irwindale Raceway, but because Caltech was in the neighborhood and he had technical talents, he gave up hot rods—except for the 1942 Jeep with a Chevy V8 engine, Turbo 400 transmission, and '56 Pontiac Positraction rear end—and became a machinist at Caltech. For a while, he worked with Jim Westphal, who then introduced him to Jim Gunn. Michael liked working with Jim Gunn: Jim appreciated his work and didn't pressure him and besides, he was a cool guy.

After Jim went to Princeton, Michael continued working for Caltech and JPL until 1991, when he heard that Jim had started a new project. He thought, "Should I call Jim?" At first he thought not—after all, he wasn't a degreed engineer—but he'd also learned something about not missing opportunities; he'd once coveted a '65 Lamborghini without ever telling the owner he wanted to buy it, and the owner had sold it to someone else. So Michael called Jim and said, "I want to come to work for you if you want me to." And Jim said, "When can you get here?" So on April 15, 1993, Michael went to Princeton, on the engineers' usual soft money, and has been there ever since.

Jim and Michael complemented each other well. Jim did the camera's preliminary design, Michael selected materials and took care of the details of the mechanisms. He believed he was a good inventor of mechanisms, one of which assured, on the first try, that all the camera's CCDs were aligned to within a thirtieth of the diameter of a human hair. "I've worked my entire career to learn to work on this camera," Michael said. Jim calls Connie "C" and Michael "Michael"—as in "Michael, could you go into the shop and make this hole smaller?"—and felt that he and Michael just clicked. Connie ended up being responsible for all the electronics, camera and noncamera, for testing, fixing, fine-tuning: at some point, she said, she knew "how to reboot every piece of electronics that ran the telescope and the instruments." Jim felt he clicked with her too. "She's just a pleasure to work with," he said, "and I think she feels the same about me. I don't know for sure but I suspect so."

Connie had graduated from Princeton in 1993, about the time Michael showed up. She went to graduate school in astronomy at the University of Chicago, where her adviser was Don York, because she'd made a conscious choice to stay near the Sloan— "because of course it was going to be finished in a year or two," she said, "so I would be around to get some data and keep my hand in." At Chicago, she did what her classes required her to, then spent the rest of her time, summers included, on the camera. By the summer of 1996, she didn't see who would finish the work if she didn't, so she spent the next year at Princeton. She assumed that Don ran interference for her with other Chicago faculty members who were wondering why she was taking so long to get through a degree. "On alternate Tuesdays we all feel sorry for Connie about delaying her PhD," said Jim, "but I'm afraid that the mercantile facts are, we can't do without her." A postdoc at the Institute for Advanced Study named David Hogg noticed that everyone in the collaboration who hadn't actually met Connie thought she was an old person, like Jim Gunn. "They cannot believe that a graduate student," he said, "a *girl*, is building the camera."

The camera was complicated in every respect. The CCDs needed not only to line up perfectly with one another but also to not move. They run best cold, around -100 degrees centigrade, and since a coolant failure would ruin them, each of the six columns of CCDs has its own dewar, a sort of thermos bottle, of coolant. The whole thing was housed in a container that provided its own environment—its own cooling system and its own vacuum system—so that no part of the camera was in contact with the outside world, except through an umbilical cord.

Turn the container and camera upside down—always obeying the sign that says, in classic Jim Gunn sentences, "Never turn the camera over unless it is known to be warm and known to be dry and has pressure on the bellows. Never. NEVER. NEVER. NEVER."—and then open the container and notice that the back of the camera is an incomprehensible but tidy mess of coils, tubes, plugs, and fibers; the recurring argument is over whether it looks more like the engine of a Japanese car or one from 1950s Detroit. Close it up and turn it right side up, open the container again, and inside are the six columns, five CCDs in each. Hovering a quarter inch over each CCD is a filter that lets in light preferentially, only in a certain band of wavelengths: the r filter was red, the i filter was infrared, the u filter was ultraviolet, the z filter was extremely red, and the g filter was green. Put together, says the entire Sloan consortium, it's an acronym for Robert Is Under Ze Gunn, meaning Robert Lupton. Looking down on the camera, the filters are squares of flat, uncannily smooth glass that give the illusion of depth and look to the eye to be variants of reds and golds and greens and blues, meticulously aligned. Mike thought it was beautiful and took hundreds of pictures of it. "It's like looking at an emperor's jewel box," he said. "No," said Connie, "it's like a Mondrian stained glass window."

Connie knew each CCD and their different personalities. Some were picky about voltages, some had low-level misbehaviors that she could tolerate as long as they didn't get worse; one chip behind a u filter would sporadically stop returning an image altogether and then in its own time, up to twenty-four hours, would come back. Connie had spent several years with her face not six inches away from that Mondrian camera window, loved wicked plumbing, and thought optics elegant and beautiful, a thing you can make almost with your bare hands and still be accurate to a fraction of a wavelength of light.

Over the top of the filters is glass so fine and nonreflective it's invisible. To clean it and the filters, Jim takes a fine brush, sweeps it delicately to pick up dust, then lifts the brush and its sweepings into a ¾-inch nozzle of a vacuum that Mike holds. They hunch over and concentrate, working in synchrony, the vacuum nozzle always waiting exactly where the brush comes up off the glass, almost like the brush is feeding the vacuum. When they finish, Jim wraps the brush in foil and gently closes the louvers that sit over all the filters. Then Mike vacuums a hard cover, fits it over the blinds that are over the glass that are over the filters, and bolts it on. The camera is ready to move. It is both heavy and delicate, and when moving it is necessary, it rides in a cart called the SDSS Turtle. Jim thought he should have done a better job of making it portable. On Monday, October 20, 1997, at four fifteen in the afternoon, the camera and its support systems left Princeton, trucked by Roberts Express; it was toasted with champagne as it drove away. It would arrive at Apache Point Wednesday morning.

Heading north out of the El Paso airport on Route 54, leaving west Texas going into southern New Mexico, the ground is just flat, an ancient rift valley. The road is dead straight, runs on rusty peach dirt through ranches where it's hard to see what there is to ranch, past detention facilities, past military reservations, right next to a train track with trains a mile long. A little under two hours down the road is Alamogordo, famous for being the town closest to the site of the first test of the atomic bomb. Stop at Alamogordo to buy food—you're going to stay on the top of a mountain with the nearest store and restaurant 15 miles away—then turn east on Route 82. Within forty-five minutes you climb 9,000 feet: up a scarp on the east side of the rift, past High Rolls with its machine shop that built pieces of the telescope, into Cloudcroft, which gives the impression of being a resort town you could get kidnapped in. Then turn back south along the ridge on a narrow switchback two-lane, State Road 6563, named for the wavelength of the hydrogen alpha line of the sun, 6563 angstroms, because the road was built to get to a solar observatory. After 15 miles, through pine woods and high meadows, turn off on Apache Point Road, pass a pond, and hit the dead end at the top.

Apache Point Observatory is a cluster of buildings surrounded on three sides by pine trees and on the fourth by air and space. "The air up here is delightful for humans," says Jim, "but all those pine scents are very sticky for a camera." The buildings are utilitarian, variants of vinyl-sided house trailers that include dorm rooms with shared bathrooms and a common kitchen because visiting astronomers stay days or weeks. Apache Point's big 3.5-meter telescope looks like a modern telescope: a cylinder topped by a boxy dome with a slit through which the telescope looks. The 2.5-meter telescope lives inside a building, called the enclosure, that looks like a child's drawing of a house, a square with a big garage door, topped with a peaky roof. It sits on a man-made ledge that's cantilevered off the side of the scarp. Stand on the ledge, and far out in the rift valley is the White Sands Missile Range; the white sands are gypsum that dusts the telescope and mirror. Stand on the ledge, on the open grating that serves for a floor, look down, and get acute paralytic acrophobia.

Inside the enclosure before the camera arrived was the telescope, or some of it because it wasn't done yet, just the metal cell that would hold the primary mirror, the forks for tilting the telescope 180 degrees, and large metal rods/braces that would hold the secondary mirror 11 feet 8 inches above the primary. When the telescope was set to observe, the enclosure would ride on rails back from the cliff, leaving the telescope perched on its ledge in the smooth, predictable laminar flow of air from the valley.

Bruce Gillespie was the operations manager at Apache Point. "Everybody who comes up there feels that it's a magical place," he says. Bruce was standing once with Ed Mannery in the pit below the 3.5-meter when it was being built, and they almost wanted to hold hands—"We're not gay," he said, "it was just that feeling." He felt the same way about the Sloan telescope, even though he thought it looked less like a telescope than a nuclear power plant steam chimney. But stand outside at twilight, look out at the orange glow on the horizon, and listen to the coyotes; he thinks it doesn't get better than that.

That Wednesday, October 22, 1997, the Roberts truck pulled in to Apache Point, and Jim, Mike, and a crew unloaded the camera and stowed it in one of the buildings, just avoiding slight snow flurries and the wandering wild turkeys at the pond. Next day, they loaded the camera onto a platform anchored on the back of a white, four-wheel-drive truck, Jim in the back supervising, and drove it maybe 300 yards, producing no more than 0.2 g's acceleration, to the telescope.

The camera was to be mounted on the back side of the primary mirror, so during installation the telescope had to be pointing straight up at its zenith, meaning the enclosure had to be rolled back and Jim, Michael Carr, and the Apache Point crew had to work in the cold, open air, crawling around on the floor under the telescope, working nearly blind and hunched over in maybe three feet of clearance, half on their backs, like overturned beetles. A couple of days later, they took the camera back off again and returned it for further refinements to a building with a clean room. Hirsh Cohen brought his wife to see it, both of them in clean-room white suits, looking down at the jeweled camera. Jim and Connie were in white suits too, and Hirsh watched Jim with his pliers and his meters, working on the camera he had built with his own hands. "Such a nice scene," Hirsh thought.

"It was not quite a 'turn it on and it works' effort," Jim wrote to the Sloanies on November 19, "but pretty close." Jim was being both picky and modest; in fact the camera worked right out of the box. Afterward Bruce Gillespie took the shipping box home, floated it on his pond, and raised baby ducks in it.

By mid-December 1997, the camera had been tested and retested, the electronics talking correctly to one another and to the software. Jim wrote another e-mail to the Sloanies, publicly thanking the Apache Point staff for its help with "our insistence on widgets delivered TOMORROW and helping us rummage through junkpiles for just one special nameless artifact, etc, etc, and humoring our paranoia about surges, power outages, and lightning." The camera wasn't entirely ready to use yet, but it was almost there, Jim said, "and we are pleased, immensely proud, and terribly tired." The UW astronomer Craig Hogan wrote back: "It's like the first cry of a newborn baby. I'm deeply moved and very relieved." Jim, Connie, and Michael Carr left Apache Point and went home for a while.

Over the next five months, the mirrors were taken out of storage, the primary mirror nestled into its cell and the secondary hung over it on braces that looked like Tinkertoys. By May 1998, they were ready for first light: the first time that light hits the primary, is bounced to the exquisitely aligned secondary, then is bounced back through a hole in the primary to the perfectly positioned camera and its waiting CCDs. The mirrors and camera were all that were necessary for first light, which was just as well, because the rest of the telescope still wasn't done.

First light at Apache Point didn't truly happen until the second night. On May 8, 1998, the alignment of the camera and mirrors needed to be tested, as did the software that told the camera what to do and the other software that took the data off the camera. On the night of May 9, the astronomers who wrote much of that software—Robert Lupton, Princeton's software genius, and Jim Annis, the Fermilab postdoc who had moved with his engineer wife and their newly born twins to Apache Point for the year—were at the computers in the observers' room. Other astronomers sat at other computers. Along the back wall sat shelves of monitors. Jim Gunn was everywhere at once: Jim Annis had heard Princeton graduate students say that their view of Jim Gunn was a back receding at the speed of light, redshifting down a hallway, but Annis's view was the opposite, Jim Gunn coming at him, blueshifted, saying, "Gentlemen? Couple more things to fix!" That night, Jim Gunn blueshifted into the observers' room and said, "Ready to go, gentlemen?"

At 8:15 p.m., the sun down, the telescope enclosure rolled back slowly, making warning beeps then a high whine until the telescope sat naked on the ledge over the cliff, then rotated up to look at the sky. The moon was out. The first thing the Sloanies had to do was what every astronomer since Galileo has done: get the telescope situated on the sky by finding what's called the meridian, a coordinate line drawn by starting with the zenith straight above Apache Point, moving to Polaris, the North Star, and then continuing that line to the north and south down toward the horizon to the celestial equator, the great circle drawn in the sky above the earth's equator. Normally these days this is done by hitting keys on a computer, summoning software that tells the telescope where all these coordinates are and how to get there. But the Sloan telescope didn't yet have either the software or hardware necessary. So the Sloanies had to find Polaris by hand.

Jim and Connie went outside to the telescope on its dark ledge with a laser level, a regular builder's level with a laser that could be aimed at 90 degrees from the level. They'd first find Polaris through the telescope, then mark on the floor with a Magic Marker which way the telescope was pointing, and that was north. Then they'd aim the laser 90 degrees away to the right, make another mark on the floor, and by hand, move the telescope right 90 degrees, and that was east. And then they'd find the next 90 degrees to get to the south and so draw the meridian.

Once they had the meridian, they could begin drift scanning: park the telescope on the celestial equator, use a regular bubble level to level the telescope, and let the stars drift past it. Those stars would drift, not through the eyepiece of the telescope, but down Robert's and Jim Annis's computer screens. So while Jim Gunn and Connie were outside aiming the telescope, Robert and Jim Annis were inside seeing what the camera saw. Annis hit the Enter key, triggering the goStare command, which told the camera to look at Polaris. Little white smears drifted down the computer screen. They weren't Polaris. Jim Annis decided they shouldn't look for a blinding star on a CCD, but for star tracks instead. Because the sky actually appears to circle Polaris—the stars nearer the celestial equator move in larger circles, the stars nearest the north move in small circles—Annis recognized that as the telescope got closer to Polaris, the little white smears would go in tighter and tighter arcs. Find the tightest arcs, and they'd find Polaris. A charmingly clever idea, but they still couldn't find the star. The camera was working, random light was hitting the CCDs, but they couldn't find Polaris.

Robert and Jim Annis typed on their computers; Don York looked over their shoulders at the screens; Jim Gunn and Connie came in and looked over their shoulders, and so did a lot of other people; then Jim went back out. Dan Long, one of the Apache Point technicians, said over the intercom to Jim, "You're moving the telescope down. We need to go up." Jim Annis told Connie that they finally knew where Polaris was but they had gone the wrong direction; Connie said, "Hffff" with a smile and walked out. Dan Long watched a computer screen: "There we go," he said. "At least half a degree. You're about a third of the way there." The phone rang, Robert answered and said, "Hi, Jim. This is

Robert. We're still attempting to find Polaris. We found it once but lost it again. Call back." Astronomers and technicians paced. An astronomer calculated, "Assuming I know what I'm doing. We want to move forty-five arc minutes in the opposite direction we moved last time, in altitude," altitude being jargon for a telescope's up and down motion, as opposed to azimuth, which is movement from side to side. Connie came inside and leaned against the desk, hands in her pockets. "Polaris is bright," she thought, "you can see it with your naked eye, and the idea that we have a two-point-five-degree field of view and a two-point-five-meter telescope and we can't find the North Star is, on the face of things, completely absurd." Jim Annis punched buttons on a calculator and thought, "We have the biggest camera on the planet, two hundred million pixels, twenty astronomers, that's a lot of PhDs, and Jim Gunn walking up and down in the control room, 'It's been four hours, gentlemen.' We're excellence incarnate. We cannot find Polaris."

Eventually they found star tracks in tight enough arcs and declared it Polaris. Jim Annis said, "Let's just go ahead. Let's pretend we got it." Connie told the intercom to the telescope, "Go ahead and mark the azimuth disk if you're ready." The intercom replied, "They're doing it now." It was 5:30 a.m. Outside, the telescope was ghostly in the white dawn. Wind blew through the Tinkertoys. Jim walked back outside. Technicians lowered the telescope by hand to a horizontal position, Jim paced around it, hunched over in the cold, watching while the enclosure rode back over it with a high whine. The sun came up.

Nobody cheered, nobody opened champagne. "You can't celebrate finding Polaris," said Jim Annis. But at the time they found it, Jim Gunn had walked over to Jim Annis and said, "Did you notice it was circling, how smoothly?" Jim Annis said he did. Jim Gunn said to him, "Brilliant!" meaning Jim Annis's software had done brilliantly. Jim Annis thought, "Jim Gunn's saying 'brilliant' to me. OK, I can die now."

By 6:00 a.m., May 10, Dan Long had written an e-mail to the collaboration, "To sum up tonight's observing in two words . . ." and then spelled out FIRST LIGHT in inch-high letters made of asterisks. Jill Knapp wrote a fancier e-mail, using asterisks not only to spell out FIRST LIGHT but also to create a bubbling champagne glass and the heading "It was a long time coming."

The following June, having improved their software and hardware to the point where the stars were no longer smears but stars, the Sloanies presented at an American Astronomical Society meeting a cleaned-up version of eight minutes' worth of one scan of the sky, 1 percent of the first night's data. They stood in a line and unrolled a print, about 3 feet high and 30 feet long, full of dots that were stars and quasars and dots embedded in soft fuzzes of different shapes that were a whole zoo of galaxies. One astronomer at the meeting thought the Sloanies were showing a computer simulation. "Oh," he said, "this is *data*?"

Jerry Ostriker was there. Earlier he'd written an e-mail of congratulations, four sentences with eleven exclamation marks, that began, "It has been a long haul. And a huge amount of hard work and smarts went into this (not to say $$ as well!)." Rich Kron continued his tradition of brewing celebratory beer and brought a case of it to the meeting. Most of his beers had labels designed by amateurs, but Fermilab—"the only place that was sufficiently organized and knew how to do these things," Rich said—asked the designer of Sloan brochures to design a label for First Light beer: "1st Lite Ale" in a classic oval printed with the SDSS logo and the motto "Brewed nightly in the mountains of New Mexico" set in a deep blue field with those same fuzzy galaxies.

It was all a great success. The Wednesday after the Sloanies' presentation, their website got 140,000 hits. CBS News covered it, it went out on the Associated Press news wire; the *New York Times* and the *Boston Globe* wrote their own articles. Most reports quoted Jim Crocker saying, "This is not just a telescope, this is a science factory."

The factory produced, and data flowed. The telescope still wasn't tracking, the spectrographs weren't on it yet, the camera still needed adjustment, the operations software needed debugging, and the photometric pipeline needed further testing, but the data rushed out of the camera onto magnetic tapes, the tapes were shipped overnight to Fermilab, where the data were sent through the photometric pipeline, and out came stars, quasars, galaxies.

On June 1, 1998, Jim Annis wrote an observers' report: "The nice cluster we saw on MJD50962, run 39, field 232 is UGC 09792, otherwise known as Palomar 5. We've rediscovered a galactic globular, and Robert gets a cigar. Hey, Connie, want some more data for your thesis?" Connie, who was in her fifth year of graduate school at Chicago, needed a thesis topic, and one of the Chicago astronomers looking out for her suggested she look at globular clusters, in particular the one called Pal 5. Globular clusters—which Jim had also studied—are dotted around the outskirts of our galaxy, perfectly round balls of some of the oldest stars in the universe. No one knew how they had formed in such a way that they all seemed to be the same age.

Connie wanted to find any stars that might have been pulled away from the Pal 5 cluster by our galaxy's gravity, with the idea that those errant stars might hold the history of the relations between the cluster and the galaxy. She also wanted to do another check of age—if the stars all formed at the same time, they'd have the same fractions of materials, the same chemical composition—but she couldn't because chemical composition could be measured only with spectrographs, and the Sloan spectrographs were still at Hopkins. So Connie worked on Pal 5's images and otherwise went back to maintaining the camera.

Data flowing didn't necessarily mean science. The hardware and software both were still too rickety for reliable accuracy and precision. Nevertheless, the following month, July 1998, Michael Strauss and his graduate student Xiaohui Fan used data taken a couple weeks after first light, partly to test the photometric pipeline, partly to look for quasars. Fan had come to Princeton with the fixed goal of finding high-redshift quasars. Because finding quasars meant using the results of the photometric pipeline, and because Robert Lupton had written it, Princetonians were what Strauss called "close to the data," meaning that Strauss and Fan could walk down the hall to Robert's office and say, "Robert, this image doesn't look kosher. Should we believe it or not?" One of the pipeline's jobs was to take the camera's stars and starlike things and plot them, on what's called a color-color diagram, by two different measures of their colors. Stars fall along a nice line from blue to red. To find the nonstars, Strauss and Fan looked for the things that fell off the line, the very reddest things, which could either be quasars at high redshifts or, Strauss said, "maybe was just one of the hundredth percent of the gazillon stars for which the software had screwed up."

By mid-July, Fan had a list of things with the colors of quasars. To verify the redshifts, they used the spectrograph on ARC's 3.5-meter telescope on Apache Point, and since the 3.5-meter is operated by remote control, Strauss and Fan were spending their nights at a computer in the basement of Peyton Hall, working their way through Fan's list. By the end of September, they found ten new quasars, the farthest with a redshift of 3.53. By mid-October they announced a quasar with a 4.75 redshift, second only to Jim's

1991 quasar with a 4.90 redshift. On Thanksgiving morning at 1:30 a.m., they found a redshift 5.0 quasar. Strauss thought, "We gotta call Jim." So somewhere around 2:00 a.m., they did. Jim, who says he never thinks about anything in the middle of the night, said something like "Very good, boys," and went back to sleep—in spite of which, Strauss thought Jim was pleased.

In December, Fan and Strauss told a collaboration meeting that they now had twelve new quasars with redshifts over 3.6, that three of the four farthest quasars known had been found by the Sloan, and that those twelve quasars had been selected from nineteen candidates, meaning their success rate was 70 percent, much higher than the 10 percent rate of surveys with photographic plates. Most importantly, the software was working. The audience burst into spontaneous applause.

Chapter 7

Drift Scanning

Saturday September 12, 1998
Apache Point Observatory SDSS 2.5m
Observing Log
Enclosure rolled off at 18:00. It was a mostly clear to clear
night with nearly calm winds. The first night back on the air
since the end of June had its expected glitches, but most of the
night was productively spent imaging Polaris.

> —Dan Long, observer

FIRST LIGHT HAD been accomplished with the least telescope possible—a camera, a primary and secondary mirror, and just enough Tinkertoy rigging to keep them all
aligned. So the next thing to do was to clothe the naked mirrors
with some structure that protected them without also degrad-
ing the sharpness of their focus, and focus is degraded by lots
of things, including wind, light, and heat. A suite of baffles was
installed around the mirrors to deflect winds that shake the telescope and to reflect out any stray light. The baffles also had to

be open enough to hold the temperature of the whole telescope, and especially of the mirrors, at the same temperature as the air at Apache Point; a room-temperature mirror out on a cold night has little heat ripples flowing over it and ruins what astronomers call "seeing." French Leger, a UW engineer, said that the point was to enclose the telescope so that it thermally disappears. The stealth telescope is impressive in its oddity; it looks like a louvered refrigerator box with petals on top, called flatfield screens, that open and close over the mirrors.

During the day, the person in charge of the telescope was French Leger. He had been an engineer working with a physicist at UW, and when the physicist's grant wasn't renewed, he got a job drafting for the UW engineers on the Sloan, who, when they were running behind and discovered he could also design, asked him to do so. Pretty soon he was commuting between Seattle and Apache Point, helping to get the telescope together. French is Canadian and grew up in Saint-Henri-de-Barachois, 5 miles due east of Shediac, New Brunswick, then moved to Goose Bay, Labrador, for high school and then to three other high schools. Because of the moving, he didn't have the credits to graduate from high school, so he didn't. Instead, he joined the U.S. Navy as a Canadian citizen, found a book on board ship by George Gamow called *One Two Three . . . Infinity*, and got excited about physics. So he started college, and after five universities, a marriage, three children, a divorce, and twenty years, he got a bachelor's degree in physics. His name is not actually French; it's Roger. His navy uniform had "Leger" written on it, and one day someone said, "You're French, right?" and he said yes. He said people occasionally called him Frenchie, but they had all died. In 1999, Crocker appointed French the Sloan's telescope engineer. What French liked best about the Sloan, he said, "is that what we're doing goes to the public and not into some black-box military b.s."

French worked with a crew of mostly local people who live not less than forty-five minutes from Apache Point; working at a telescope is a good job in that part of the country. French's crew is in charge of the telescope during the day. They're alert and talkative: every action is a discussion about what's to be done in what order. The size of the telescope doesn't leave much room inside the enclosure, and they carry three-dimensional mental maps of all the wires, equipment, and shelves and never run into things. They climb a lot, hang on to the edges of things, and are aware of their jobs' dangers; one of them slips on the metal stairs and everybody jumps and runs. "If someone gets hurt," they say, "and it can't be fixed with a Leatherman and duct tape, we just roll 'em off the front of the telescope and let the coyotes get 'em."

At night, the people in charge of the telescope were the observers. They were usually PhD astronomers who'd signed up to work on the mountain for five years; they were occasionally spelled by visiting postdocs. They worked two per night and made all the nightly decisions about whether to observe, when to open up, whether to use the camera or spectrographs, what parts of the sky to point to, when to close. They were working with hardware and software that were unique, complex, required a lot of learning, and that either broke or were being almost constantly improved.

They'd run through their preflight checklists: check the photometric telescope, check the 2.5-meter telescope, check the telescope drives, check the weather. If it was heavily cloudy, raining, snowing; or blowing around too much pollen or the white sands of the White Sands Missile Range; or so humid that moisture condensed on the mirrors, they wouldn't open the telescope. If conditions were better, they'd open, watch their banks of computers that monitored whatever on the telescope needed fixing that night, check the weather again, monitor the seeing, and write the night log, working until morning twilight. They'd stow both telescopes, do a walk-around, say good morning to French, then go home to bed, thinking about what they'd done that night and planning for the next night and how to do it better. The observers were responsible for the Sloan data's uniformity and were a little irritated at working so hard at something so crucial and complex that left no time for their own research: "Quite frankly," said Stephanie Snedden, who eventually became lead observer, "we do a hell of a lot of work for very little recognition." The deputy lead observer called himself the deputy lead assistant plankton.

They worked in a dark room, lit only by computer monitors and a lava lamp. Astronomers have always liked to enliven darkened rooms in the middle of the night with quiet, moving light, usually from a fish tank, but an early Sloanie had brought the lava lamp to Apache Point because his wife had given it to him for luck and the observers needed the luck. The observers noticed that if the lava lamp wasn't turned on at the beginning of the night, something broke, and they began including "lava lamp on" in their preflight checklist. Night log #435: "Very, very bad seeing (averaging 4 arcsec) to start. We attribute this to the lack of thermal management provided by the lava lamp."

On a remote mountain working the night shift, the observers' night logs got goofy. A program called endNight—a long list to be checked off before shutting down for the night, including ensuring that the data was now on tape—could never be gotten to work right: "It never worked. It never worked, it never worked," explained Jim Annis. "It was so painful and took so long, and in the end you get nothing out of it except you get to go to bed." One night it worked, and thereafter, the observers, remembering the Monty Python movie with the line "and there was much rejoicing" accompanied by an unconvinced "yay," ended the night logs with "endNight rejoiced" and implied the "yay."

Finally the night logs started including holiday menus:

Egg rolls
Squash soup
Fresh green salad
Fruit salad ala French
Roasted Turkey and Gravy
Fresh Cranberry sauce
Sourdough Stuffing (what other kind is there?)
Mashed potatoes
Sweet potato and pecan casserole
Cranberry bread
Nut bread
Christmas Stoellen
Polish pancakes
Apple sparkling drink
Pear sparkling cider
Hot Snedden apple cider
Pumpkin pie with whipped cream
Carrot cake
Kenyan coffee
Christmas cookies from Jackie (thanks Jackie!)
Brownies from Craig who shoveled snow off Russet's
house and lit the furnace, then gave her brownies.

One item on the observers' checklist was to inspect for moths. Apache Point called them miller moths—later research revealed them to be the adult form of the black cutworm—and every spring in New Mexico, they migrate to high elevations. During a bad year, Apache Point buildings could have hundreds of moths a day.

Moths like dark, tight spaces, including the insides of electronic devices and the drives of telescopes. When the telescope moves, the moths are crushed on the drive surfaces, and the telescope slips or loses track of where it is. So the telescope drives had to be cleaned during times when observers had better things to do. "The moths cause us all kinds of hell," said French.

A college student interning with the UW engineers was commissioned to investigate moth-ejection measures. The moths were most active during hailstorms and seemed agitated by jingling keys. So the UW student, who had a scientific bent, smacked the walls and floor, then stomped and clapped, but the moths showed no sign of being annoyed until the racket was a few inches away. Then he aimed a speaker at a group of moths and broadcast sounds in a wide range of frequencies. At 0.4 kilohertz, the moths startled but settled back down; at 0.6 kilohertz, one moth moved briefly; at all other frequencies, the moths remained unconcerned. Then he tried shining beams of both incandescent and fluorescent light by turns on the group, and they all moved to the edge of the light. He turned the light off and back on again and forgot it until the next morning, when he found that the moths hadn't moved at all. Next, following advice from Apache Point veterans that blowing air seems to profoundly disturb moths, he proposed what he called a conceptual deterrent system that involved an air gun, small rubber tubing, a valve, and an electric timer. He built the system and blew jets of air at the moths; they tended to hunker down and wait until the jet was over. He was impressed that they could handle upward of 80 pounds per square inch, though they became restless with as little as 20. He installed this air-jet blowing system on the telescope and while it was working, no moths were run over— implying, he thought, some degree of success.

Further research found that, as French said, "interrupted air blasts they do not like." Eventually he built a moth ejector system that blew air intermittently through pipes with holes along their sides: "A 2 hertz puffer system is what we found to be most effective." He mounted the system on the most sensitive parts of the telescope, and though the moth problem didn't completely resolve, it was much improved.

The telescope's real problem was the one that had made first light so nearly frustrating: it couldn't point. The observers would type in the coordinates on the sky and hit Enter, and the telescope was supposed to go to those coordinates immediately. Night log May 29, 1998: "Another clear night. We aimed to fill in the stripe of the data we took last night. In order to do this, we had to obtain the current position of the telescope. We got on the sky at 9:10 p.m.; we had a position at 11:15 p.m."

Or they'd try to find a kind of star with standard and known brightness, called an FK5 star. Night log May 16, 1999: "We were unable to acquire our field. We were then unable to acquire an FK5 star. We were then unable to acquire Arcturus." The pointing was 3 arc minutes off, a tenth of a moon off, and the telescope was aiming at something that was an arc second across. The observers that night were Chicago and Princeton postdocs, and the Chicagoan, Scott Burles, wrote the night log: "At 1 a.m., we found Arcturus and returned to our FK5 star. We chased the fast and wily FK5 star for several hours." "Fast and wily" is a joke; no FK5 star is anything but steady and predictable. "Can't wait to get up in six hours and do this all over again," he wrote.

Part of the pointing problems were, strangely, because the telescope thought it was in the wrong place. The telescope has to know where it is on the earth: at the North Pole, Polaris is straight overhead, but at Apache Point, Polaris is 57.2 degrees off north, and the telescope didn't seem to know that. The problem was that the telescope had been installed with the slightest tilt, about one millimeter off perpendicular. That one millimeter on earth, translated to the sky, meant that when the telescope thought it was looking straight up at Polaris, it was actually looking 1 arc minute off to the side. One minute of arc translated to the 8,000-mile-diameter earth, Jim figured, meant the telescope thought it was living about a mile away from Apache Point. Jim thought the sensible solution was to just lie to the telescope, to tell it that it was a mile west, out in the rift valley. The UW engineers thought this was a jaw-droppingly brilliant solution.

Pointing was one of only two things the telescope had to do; the other thing was tracking, and it couldn't track either. The advantage of drift scanning is that the telescope sits still and the earth turns it across the sky with a steadiness and smoothness no telescope controls could hope for. The disadvantage is that drift scanning works effortlessly only when the telescope aims at the celestial equator, the great circle drawn on the sky directly over the earth's equator, where the stars move in long arcs that look like straight lines and slip evenly across the CCDs. But aim the telescope up off the celestial equator, and the stars move in increasingly tight arcs until, near Polaris, they are moving in small circles. So to get the star moving evenly across the CCD, the telescope needs to track, to move automatically to counter the earth's motion.

A year after first light, in May 1999, the observers were still reporting they couldn't get the telescope to track. It would refuse to move at all, it would move at inconstant speeds, it would oscillate violently, it would start to move and then slip its leash and run away to the zenith. One night that May, Dan Long—one of the few observers who was not a PhD astronomer, just experienced and natively smart—was out on the cliff with the telescope and later wrote up a report: "I heard the sound the alt lvdt makes when the windscreen changes position with respect to the altitude. I looked up with a flashlight to see oscillation in altitude. By the time I got to a stop button, the oscillations were so violent that the turnbuckle rods in the windscreen were slapping the sides of the windscreen making a loud clang. At least that is what I hope the noise was and not the telescope hitting the windscreen."

As a result of the telescope's bad behavior, most of the data during that first year came from drift scans at the celestial equator —nothing was wrong with that except the data were all coming from a strip near the horizon, and the Sloan was supposed to be surveying the whole sky. Many of these problems were in the software, much of which had been handed off from the UW engineers to Fermilab, whose software writers, said Jim Gunn, "had never seen a telescope." So Princeton took over the software: Robert Lupton, whose PhD thesis had been the code that analyzed the data from PFUEI and Four-Shooter, went to Apache Point for three months and rewrote the code until it worked.

So far, the problems had been more or less as expected. But on October 19, 1999, an Apache Point engineer named Jon Brinkmann was showing another engineer, John Briggs, where on the telescope the spectrographs were mounted. Briggs looked up through the open mounting hole and said, "Jon, is that a crack?" Brinkmann and Briggs moved their heads back and forth, to see if it was just a trick of lighting; it wasn't. Brinkmann called French. "French," he said, "do you know there's a crack in the secondary?"

"Is that really a crack," French said, "or is there something on the mirror?" So they crawled up inside the Tinkertoys to the mirror, and sure enough, it was a crack. French walked over to the operations building to get Bruce Gillespie and brought him back.

"See anything weird about that mirror?" Briggs said to Gillespie. "Look at the center."

"Shit," Gillespie said. "Oh shit."

This was scary: mirrors are under high internal stress, and cracks propagate. All work stopped, nobody touched anything. French and Gillespie called Jim Gunn and then Fermilab's Bill Boroski, the project manager who had just replaced Jim Crocker. That was early afternoon, and within a few hours everybody was discussing options on phonecons. French thought it might be a project killer and worried about people's jobs: fixing the mirror could cost half a million dollars and take a year or two, and French had just moved with his wife from Seattle to Apache Point, and newly hired observers had been buying houses nearby.

Jim and Boroski both flew out the next morning; glass experts from Arizona's Mirror Lab came to inspect. The mirror had cracked in a meandering circle around its center, the crack stopping an inch or so before closing on itself. "Oh, man," French thought, "it could shatter the whole mirror, it could fall into a million pieces. You just don't know what's going to happen, mirrors are very unpredictable." Jim, French, and another new UW engineer, Larry Carey, covered the cracked center with foam and tape, then gingerly took the mirror off the telescope and moved it into the support building. The Mirror Lab experts drilled tiny holes at the ends of the crack, a temporary fix that stopped the crack from continuing.

Jim flew to the Mirror Lab and everyone discussed what to do. They examined the mirroring surface and found it had not distorted. The glass was under too much stress to just squirt cement into the crack. But because the primary mirror has a hole in its center through which light passes on its way back from the secondary, the center of the secondary doesn't receive much light. So French drove the mirror to the Mirror Lab, the Mirror Lab cut out the crack, leaving a hole in the center, and capped the hole. French's worries had been pessimistic: the fix cost only around $50,000, and the mirror came back to Apache Point in three months, the following January, the beginning of the new millennium.

Meanwhile, Jim rechecked the design he'd done on the computer for the mirror's controls and found he'd made some typos that caused one little rod to be a little too long, so every time the mirror moved, it hit the rod and eventually broke. "There was no problem with the secondary until I broke it," Jim said. "Most decidedly it was my mistake." Gillespie thought it was Jim's dark moment: "Nobody yelled at anybody about it, though," he said.

Michael Strauss thought the project had just about used up most of its nine lives. The Sloanies all knew they were running on borrowed money, and they thought, as Jill Knapp said, that their sponsors' reaction would be that the broken mirror was "the last damn straw—these people can't do anything right and now they've gone and broken their mirror. Forget it." That the sponsors worked out a way to stay in the game, Jill thought, was partly because of Strauss's and Fan's quasars. "This beautiful science—it was like lifting up the corner of the tent and getting a look at what was inside, and the world could see it. And that got us through, I think." A month later, on January 26, 2000, the fixed secondary was reinstalled on the telescope, and it worked just fine. The observers went back to battling moths and white sands, rebooting software, and looking for sucker holes in the cloud cover.

By this time the various pointing and tracking problems had mostly been hunted down and eliminated. The telescope could point to anywhere on the sky and be where it was supposed to be. And just to make a thoroughly cozy transition out of the twentieth century and into the twenty-first, on December 30, 1999, John

Peoples, who was now Sloan's director, wrote an e-mail to all the Sloanies, telling them that the Sloan Foundation had just given them a new grant. Peoples, talking to Hirsh Cohen, had asked the foundation for $6 million. Hirsh Cohen had been sympathetic: "They're really stuck, and I want to keep them going." He wanted to give the survey the $6 million requested, but first he talked to the foundation's director, Ralph Gomory, about whether to stop funding the survey altogether, having already sunk $10 million into it. And Gomory said no, they shouldn't fool around, give them another $10 million. By now, Hirsh, trained as an applied mathematician, had gotten to know the Sloanies and begun to understand the science factory: "It was just beginning to dawn on me that we had a new kind of scientific machine," Hirsh said, "that you poured light into it and out came stars, quasars, and galaxies. I became part of their crowd."

Chapter 8

Spectroscopic War

The word "idiot" gets used a lot. Astronomers like that word.
—Michael Strauss, Princeton University

A S OF THE end of 1999, the telescope was pointing and
tracking well enough, Jim's camera worked like a dream,
and Robert Lupton's photometric pipeline, though still
buggy, was running robustly. It was processing one night's worth
of objects in two days, which was adequate, and identifying, locat-
ing, measuring, and characterizing those images that should be
revisited for spectroscopy. Spectroscopy couldn't be done yet,
because the spectroscopic system wasn't working.

Alan Uomoto and Steve Smee at Hopkins had finished the
spectrographs and in the spring of 1999 had installed them on the
telescope. Hopkins was pleased with itself that though it ran late,
it was on budget and more nearly on time than its sister institutions; it liked not having to apologize. Once the spectrographs
were installed, Uomoto moved on to other projects, and Smee—
being a barely trained, soft-money engineer no longer getting

paid—went off to graduate school to get a PhD in mechanical engineering. When Uomoto and Smee left, however, the spectrographs still hadn't been tested on the telescope. The job was assigned to Bob Nichol, who'd worked on the survey as a postdoc at Chicago and was now a young assistant professor at Carnegie Mellon University.

In one of Nichol's courses at Carnegie Mellon were three undergraduate juniors who were energetic and smart enough that they were going to graduate early; in the meantime, they were bored. Nichol thought they knew more about computers than he did, so he got permission to hire them to help with the spectroscopic commissioning—that is, with getting the spectrographs' several subsystems, which included software and had been spread out among the institutions, to work as one system.

The Sloan had two spectrographs mounted behind the primary mirror on either side of the camera; the camera can come and go, but the spectrographs stay put. Like the camera, the spectrographs worked beautifully pretty much from the start. On May 30, 1999, Nichol sent the Sloanies an e-mail headed "Spectra!" saying they'd gotten a galaxy redshift of 0.07, meaning the spectrographs had passed their first test and now had only 999,999 more redshifts to go.

But the spectrographs themselves were only one step in getting spectra into an archive, and the whole spectroscopic system was a complicated little factory. Once the camera collected the pictures and the pictures were recorded on tape, the tape was sent via Federal Express to Fermilab, where its content was put into the photometric pipeline, which fed into the target selection pipeline, which decided what objects were interesting enough to have their spectra taken. The list of interesting objects was sent to UW, where their positions were plotted on a map, which went to room-sized drill press, which drilled the map, one eighth-inch hole per galaxy, 640 holes in all, in a round aluminum plate 30.3 inches across, corresponding to three degrees of sky, six moons across. A finished plate looked like a large aluminum pizza pan with constellations drilled into it.

The plate was then sent in a boxed set of ten plates back to Apache Point. At Apache Point, next to the telescope enclosure, in a little building called the plug lab, a plate is held by its edges in a frame called a cartridge. Every day, two Apache Point employees called pluggers lean over a plate and in about forty-five minutes plug into each of those 640 holes one end of an optical fiber, which slips into the hole with a satisfying click. The other ends of the optical fibers are gathered, twenty at a time, and clamped into thirty-two blocks. And then the cartridge, plugged plate, and blocks holding 640 fibers altogether—the whole affair now looking like a framed pizza pan with its hair sticking straight up—is carted out to the telescope and plugged into the spectrographs.

At twilight, the observers begin looking at the weather. If it's clear, dry, and still, they decide to use the camera. If it's slightly cloudy, they'll use the spectrographs, whose CCDs—unlike the drift-scanning camera's—can be exposed longer. And if it's clear and then slightly cloudy, they'll use the camera first and then the spectrographs. The camera/spectrographs swap is unusual; most telescopes devote the entire night to one instrument, and any night that's too bad to use the instrument is also just too bad for the astronomer.

On spectroscopic nights, the enclosure rolls back, the telescope opens up and points at the relevant patch of sky. Light from each one of 640 targeted objects hits its exact hole in the plugged plate and runs down its own fiber to the spectrographs, which turn out 640 tiny rainbows, captured on CCDs and recorded on tape. On a good nine-hour, nine-plate winter night, the observers could get the spectra of up to 5,760 galaxies—"A very, very high production machine," French called it.

By the end of June 1999, Bob Nichol reported to the Sloanies that on one night, they'd taken more than five hundred good spectra. By November, they'd taken ten thousand spectra total. Meanwhile, Nichol was burning out—assistant professors need to worry about doing famous science so they can get tenure, and Nichol had a wife and new son on the way in Pittsburgh while he shuttled back and forth to Apache Point. He left the mountain for good and went back to Carnegie Mellon to do his science. Besides, he said, the only nonastronomical things to do while living in Apache Point were go to bars or ride the miniature trains in Alamogordo, and he didn't have time to do either anyway. He was proud that he and his undergraduates had gotten the spectroscopic commissioning kick-started, but he knew the spectroscopic system was still held together by string; it was creaking, it wasn't tuned.

While Nichol was still at Apache Point, he had had the sense that Princeton wasn't quite trusting him to do the spectroscopic commissioning because Jim Gunn had sent out his postdoc David Schlegel, who had no other obvious reason to be there. When Nichol asked Schlegel why he'd come, Schlegel said, "I'm here to help." At the time, Nichol thought Schlegel's attitude was that he, Nichol, was an idiot.

Schlegel said Jim had sent him to Apache Point, but he didn't know why—maybe because Jim wanted someone he knew to check out the whole spectroscopic system. Jim was in fact worried that the spectroscopic system wasn't ready for business but he wasn't sure whether he'd sent Schlegel or not.

———————

David Schlegel was at Princeton in the first place partly because he was an expert on dust. Dust is an astronomical nuisance. Astronomers' only evidence of the universe comes through light, but dust dims things that shine so you don't know how bright they really are, or it hides them completely. Dust absorbs blue light, making anything seen through it look redder; it gets heated by nearby stars, then reradiates the heat as infrared light or microwaves. It differs in different parts of the sky, clumping up in some places, spreading out in others; in still others it is made of different particles that affect light differently. Since the Milky Way is full of dust, and since we live in the Milky Way and see the rest of the universe through it, all our observations are intrinsically untrustworthy.

When Schlegel was still in graduate school, he and his adviser, Marc Davis, and another graduate student named Doug Finkbeiner mapped the dust and published a highly technical and eventually highly cited paper on how to correct for it. After Schlegel finished his PhD, he took a postdoc at Princeton to work with Jim Gunn on the Sloan; when the Sloanies needed to write the pipelines to correct for dust, Schlegel knew all about it. He wrote the dust correction code quickly and then moved on to unrelated research, waiting for the Sloan data to start flowing so he could do some science with it and get a real job.

Schlegel has a physical presence that's part watchful intensity, part gawkiness, part charm. He has a tight, close-mouthed smile; his voice is slow and sleepy. He was young, not yet thirty, and he disliked authority. His friend David Hogg told him he'd never make it in the military, and he took that as a compliment. He was deeply interested in the survey's success, was unusually good at writing code, and was spending 200 percent—his figure—of his time doing it.

While waiting for data, he began showing up for Jill's photometric pipeline lunches, heard the Princeton people worrying about the spectroscopic system, and wondered why a system booked for a large fraction of telescope time was still being worried about. Schlegel knew about spectrographs from other projects he'd done, and so in the spring of 1999, whether Jim sent him or not, he'd gone to Apache Point. He had no assignment, nothing particular to do other than talk to people, crawl around on the telescope—"I'm curious," he says—and have fun. Apache Point's biggest trouble, he found, was an ingenious part of the spectroscopic system called the fiber mapper.

The pluggers, sitting in front of their plates, do not put fiber #1 into hole #1, which corresponds to galaxy #1; at 640 fibers per plate, that would take all day. Instead, they plug random fibers into random holes. Then the plates are sent to the fiber mapper system. The system was Jill Knapp's idea: you don't care which fiber goes into which hole; you care only about what comes out the fiber's other, unplugged end. And since the other, unplugged ends of the fibers are all lined up and set into a block, you can shine at the block's end a laser that runs down the length of the block, sending light back down the fibers one by one and out the holes in the plate. Sitting and looking at the plugged plate, the first light you see is from hole #1, therefore fiber #1, therefore galaxy #1; the second light is from hole #2, and so on. The lights go off like little sequential fireworks and are captured by a CCD video camera. The video camera's CCDs are polled by software that matches the fiber to the hole to the galaxy. Jill thought the concept of the fiber mapper was easy, which says something about astronomers' reference frames for "easy."

The fiber mapper had been built at Fermilab, installed at Apache Point, and once the spectrographs were on the telescope, had gone into action. Fermilab thought it worked well, or at least well enough to leave alone for a while. Schlegel thought it was slow: mapping one plate took, say, three hours, meaning that a nine-plate night would take twenty-seven hours to map, and the spectroscopic data were supposed to be leaving Apache Point the day they were taken. Furthermore, the fiber mapper was mapping only about 70 percent of the fibers. The science requirements allowed the fiber mapper to miss 1 fiber in 100,000, not 3 in 10. Schlegel was irate—his word—because it didn't work and it should not have been that hard a problem to solve.

Along with Schlegel at Apache Point was a Chicago postdoc named Scott Burles. Burles' job didn't require trips to Apache Point either; he was supposed to be writing Chicago's spectro pipelines, but Bob Nichol had asked for his help at Apache Point. Schlegel and Burles together tried to figure out the mapper's missing 30 percent. Shine the laser down those fibers, and the light comes out of the holes in the plate in nearly straight beams. Look straight into a hole and down a fiber, the fiber looks lit. Look into a hole at a fiber from an angle, the fiber looks dark. They reasoned that the camera was seeing all the fibers on the rim of the plate at an angle, so all those fibers looked dark. To see all the fibers nearly straight on, the camera would have to be farther away; it was just too close to the plate now. Schlegel and Burles figured this out in a few minutes. The next problem was to figure out how far back the camera needed to be: it was currently mounted about five feet away from the plug plate. So the two of them sat in front of the lit-up plug plate and then moved their chairs back 8, then 10, then 15 feet, until they could see all the fibers shining all over the plate. Around 15 feet seemed right. Jim happened to come out to Apache Point about then, and Schlegel and Burles told him about their discovery. Jim took several minutes to calculate the optics and came up with 15 feet. Schlegel wrote an e-mail to the collaboration team explaining all this and recommending the camera be moved; "This change is mandatory," he wrote, though "mandatory" might be a bossy word for a postdoc to use. Schlegel added that he had written a small program, called evilmap, to drive the laser; to run the program, you typed "thoroughlyevil."

But mapping one plate still took too long. Schlegel, back at Princeton now, got more irate, and he said so on the phone to Doug Finkbeiner, with whom he had done the dust maps at Berkeley and who was now a postdoc there. "Well, that's ridiculous," said Finkbeiner. "That should be easy." Schlegel said, OK, fine, meet me there next month. So on August 15, 1999, they and Scott Burles met at the El Paso airport and drove up to Apache Point together. As always, since Apache Point had no food within 15 miles, they stopped in Alamogordo to get groceries, then ate at a burrito place named Maria's because Jim always ate there, and whatever Jim did, they did. When they got to Apache Point, even though it was late and they'd been traveling for hours, they met one of Bob Nichol's undergraduates named Matt Newcomb and everybody—again like Jim—got right to work.

This time they figured out the reason a plate took too long to map was that the video camera wasn't working fast enough. They calculated the rate at which the laser scanned the fibers and from that concluded that a plate should be mappable in minutes, not hours. The next morning they ordered a new video card that would make the video camera go faster, and when it came a few days later they installed it successfully, even though the manual was only online and only in German. Luckily, Finkbeiner had an undergraduate degree in German literature.

By the time they left, the fiber mapping was taking 30 minutes and making no mistakes except when moths got into the mapper. On August 19, Schlegel again wrote an e-mail to the Sloanies: "I'd like to announce that the spectroscopic fiber mapper is now FULLY OPERATIONAL. Our secret weapon was a guest appearance by Doug Finkbeiner at APO who battled fiercely for four days straight against the Mapper Forces of Evil. Additional firepower supplied by Matt Newcomb, Scott Burles & myself." He explained further that they wanted to scan each plate twice, to protect against "the famous APO moths, which are optically thick," and that they intended to be able to do this double scan in under 15 minutes.

On August 31, Schlegel e-mailed Finkbeiner that Jim had just invited him (Schlegel) to Apache Point again. "Inviting," they both knew, was as close as Jim got to ordering someone to do something. Finkbeiner said he personally was going on vacation and wished Schlegel luck. This time, to get the fiber mapper to meet the goal of a double scan in 15 minutes, Schlegel rewrote the Fermilab software matching the fibers to the holes to the galaxies. After working late again, he got the time down to a single plate in 5 minutes, 50 seconds. Entirely pleased with himself, he sent another email at 3:42 a.m., headed "Fiber Mapper Declared 8th Wonder of the World" and beginning: "French: 'Holy Mother of SDSS! I've never witnessed anything like it!' Larry Carey: 'I'm quitting my job at Washington if they don't allow me to transfer to Apache Pt. where one can witness the Mighty Mapper every day.' Bob Nichol: 'I now believe in miracles.'"

After Schlegel sent the e-mail, he stretched out on the couch in the observers' building and fell asleep. He woke later that morning; by chance, John Peoples was just walking into the building. They'd never met, so Schlegel introduced himself. Peoples recognized the name because he'd just happened to read Schlegel's 8th Wonder of the World e-mail. Peoples was now the Sloan's director. He knew the mapper wasn't yet perfect, but it was Fermilab's job; as for the mapper software, it worked well enough for the present. His own directorial problem was to keep the survey alive by sticking to the work breakdown structure. The Sloan had a stated set of priorities, and any system that was working well enough did not have a high priority. Princeton tended to be bad at priorities, got worked up about things that weren't that important. Peoples was abrupt. Would Schlegel just please stick to the things that Princeton was supposed to do?

Schlegel, who'd been hoping for praise, didn't answer. Peoples was the survey's director, Schlegel was a postdoc. He'd spent a lot of energy in a short amount of time on something he thought should be done and had been proud of the results. "This was discouraging," he thought. It was a terrible sign, if this is how you're treated for fixing something. He wouldn't treat anyone like that. Deflated, he left for an early flight out of El Paso.

When he got back to Princeton, he called Finkbeiner and together they tried to figure out why Peoples didn't know the mapper hadn't been working; to them, "not working well enough" was equivalent to "not working." Schlegel had thought his e-mail was innocent, that he was saying only, "Hey, this thing works, isn't that great?" But after he thought about it for a while, he could see that maybe his e-mail could be read as, "Hey you Fermilab idiots, you didn't get this thing working in five years." Neither he nor Finkbeiner understood clearly that the mapper was Fermilab's; both felt free to tinker with it. In fact, though Schlegel was paid to work on the Sloan, Finkbeiner wasn't paid by the project at all: his adviser back at Berkeley thought the Sloan would be a good learning experience, and he thought of himself as a free agent working on behalf of astronomy. Both Finkbeiner and Schlegel had known that Jim was worried about the mapper, and besides, regardless of what Fermilab thought, they didn't believe 70 percent was good enough. Anyway, Peoples was only the director, and directors come and go; the chief scientist, Jim, was the person they really wanted to please.

Peoples didn't think he'd been harsh with Schlegel; the

exchange had lasted only about a minute and afterward he hadn't thought it was any big deal. He knew he was sometimes harsh and sometimes impatient, but he knew too that he was a good manager and could keep to a budget, and he was extremely aware that if the Sloan didn't keep to a budget, its funders would back out and it would cease to exist. He thought Schlegel had probably improved the mapper, but he thought that improving the mapper was polishing brass knobs.

Schlegel certainly had had no intention of making anyone mad, but this attitude, he thought, would be the end of collegiality. Jill, when she heard about it, thought so too: "The trenches were dug," she thought. "The war has started."

And the war did start. The same spring Schlegel and Burles fixed the fiber mapper, they also rewrote one part of the spectroscopic pipeline, called spectro-2D, and claimed their rewritten version worked better than the current version, which was Chicago's responsibility. The person in charge of the spectroscopic pipeline, Chicago's Josh Frieman, sensibly held a contest between the two versions and decided, yes, the Schlegel/Burles version worked better. Spectro-2D was reassigned to Schlegel and Burles. Even though Burles was from Chicago, the perception among the Sloanies was that once again, as with the photometric pipeline, someone else's job had been handed over to Princeton.

Then Schlegel got worried that the other part of the spectroscopic pipeline, spectro-1D, the part Chicago still owned, was in danger of getting redshifts wrong: "There is," he said to himself, "such a thing as a right answer." By this time, he was increasingly irate about the parts of the survey that weren't working optimally and were in fact, he thought, working pessimally. So he rewrote the spectro-1D part of the pipeline too and presented the case for the superiority of his spectro-1D over Chicago's. This time, he did it publicly, in September of 2000, at a meeting in Baltimore in front of the whole collaboration: here's the Chicago pipeline's spectrum with the wrong redshift, here's the Princeton spectrum with the right redshift; Chicago, wrong, Princeton right; examples of fourteen spectra, going on long after the point had been made.

A Chicago postdoc named Mark SubbaRao, who had spent the past two years on spectro 1D, thought Schlegel was infuriating and walked out quietly. When he got into the lobby, he slammed his conference program against the wall, then left the building, found a friend from his graduate student days at Hopkins, and went to an Orioles game. The Orioles, in the early days of a long losing streak, lost.

To make matters worse, this time Schlegel was wrong: the Chicago pipeline was fine, its redshifts were correct. He'd based his comparison on one of their earlier, buggier versions and had publicly disparaged the Chicago postdocs responsible for it for nothing. Uproar followed. Chicago was outraged and wanted Schlegel held accountable. Princeton was angry at Schlegel but said accountability was Princeton's business. Things got complicated. John Peoples tried negotiating the situation but got nowhere, and when Princeton proposed an addition to the survey, he killed it off with his usual abruptness. Princeton didn't care that the addition was killed but was furious at Peoples's treatment of them. The institutional character assassinations solidified: Fermilab is intransigent, Chicago is disengaged, Princeton is arrogant.

Finally after intense and confidential negotiations on a high level, everyone backed off. John Peoples wrote that he deeply appreciated the entire Princeton software group for their many vital contributions and looked forward to working with them in the future; and that after all, everyone was fully committed to the same goals. Nevertheless, he kept on his desk, visibly, a novel by J. P. Donleavy titled *Wrong Information Is Being Given Out at Princeton*.

Fortuitously, the Chicago and Schlegel versions of the spectro-1D pipeline, unlike their authors, worked together surprisingly well. Michael Strauss found that the pipelines agreed on the right redshifts 98 percent of the time. Of the 2 percent disagreements, half were over data that was wonkety or from particularly faint objects; the other half were over objects that were, as Strauss says, "wonderfully weird." Filtering for those 1 percents turned out to be a good way to find white dwarf stars with unusual atmospheres or under the influence of strong magnetic fields or the smallish fraction of quasars called BAL quasars whose spectra were confused by hot, fast gas pouring out of them. "I think the Sloan spectroscopic pipeline is the lowest failure rate that any automated data analysis system had ever seen," said David Hogg, an Institute postdoc. "There's just essentially no failures. Nobody has that."

At the turn of the millennium, two years after the camera's first light, one year after the spectrographs were added, in the midst of internecine warfare, the survey officially began. It was operating in fits and starts—parts of the telescope kept breaking, observers ran into problems, the weather was awful—but the data from the camera and spectrographs ran through the photometric and spectroscopic pipelines and out the other end. The pipelines were operating in fits and starts too, and the Sloanies couldn't release the data publicly until they got the pipelines right. The best way to test buggy pipelines was with real data—it was the way, after all, that Michael Strauss and Xiaohui Fan had found all those high-redshift quasars.

So the Sloanies took that first, slightly unreliable data and did Sloan's first science. Pent up scientifically for years, they hit the ground running. Within a year, a whole flock of papers—around twenty—came out on widely disparate subjects, as though the Sloan was now Herschel's luxuriant garden in which children race around, picking pretty things. "Like as an undergrad I had five stars, they were my five stars, I knew all their names and everything," said a UW graduate student named John Bochanski. "And now I have fifty thousand. Or three million."

Željko Ivezić, a Princeton postdoc who had worked with Robert Lupton on the photometric pipeline, along with several others found ten thousand asteroids circling the sun in a belt between Mars and Jupiter. Astronomers had already known that asteroids travel in two separate orbits, one closer to the sun and the other 40 million miles farther out, so they reasoned that asteroids must belong to two general families. The Sloanies found that, unexpectedly, each family had a different composition, meaning that the two families weren't just one single family that split up. Instead, they would have been born separately in the early solar system, when a cloud of gas and dust condensed toward its own center into the sun, swirled into a circling disk, and collected itself into bodies of all sizes: planets, moons, comets, asteroids. Two of these bodies, rocky ones, must have in the general chaos broken apart, each into its own family of asteroids. And the asteroids, since they'd been born together, continued to circle the sun together.

The Sloanies also calculated the orbits of all those asteroids and found they were four times less likely to smack into the earth than asteroid hunters had previously thought. Then Sloanies found that a large fraction of the asteroids observed by the asteroid hunters had been assigned the wrong brightness and therefore the wrong size. After the asteroid hunters got over their pique, they began downloading Sloan data.

In the same early data, Michael Strauss and Xiaohui Fan, continuing to look on their color-color charts for unusually red things that fell off the line and might be quasars, found some things that were redder than red and that looked like a star called a brown dwarf that's invisible except in infrared wavelengths. Brown dwarfs were so rare as to be theoretical, predicted but hardly seen, so Jill Knapp called an ex-colleague at Caltech with access to an infrared telescope to request confirmation.

Stars range from hot to cool—the sun is somewhere in the middle—and the coolest are M stars, around 3,500 degrees Kelvin or less. The Sloan brown dwarfs turned out to be even cooler, ranging from 2,200 K down to 800 K—a hot oven is 500 K. They're hardly stars at all. They're smaller than the sun, but they're not planets because they're not necessarily orbiting a star. The reddest and coldest brown dwarfs have atmospheres of methane and are called methane dwarfs. They'd been predicted in 1958 and one was actually seen in 1995, but it was going around the star Gliese 229 and could be suspected to be a mere planet. Sloanies found two methane dwarfs right away, both unattached and isolated in space.

Jill Knapp said that when the news of the methane dwarf first broke, she got two hundred e-mails from all over the world, like drums on mountaintops saying, Jill. Jill. Jill. "No sooner had we done this work," she said, "than there came a cry from the south, from Hopkins, saying, 'Hey, we're interested in that too.' Then we heard a cry from Washington saying, 'Hey, I'm interested in that too.' We have so much data, right, the more people the merrier." So a loose confederation of methane/brown dwarf people set up a long list of things to do that included finding more brown dwarfs; by 2005 they had found seventy of them.

Brown dwarfs turn out to be one of the most plentiful of the Milky Way's objects—they'd seemed rare only because they were so faint. They're half-way objects, near-suns: the clouds of gas and dust from which they condensed were too small to heat themselves to the point of thermonuclear ignition. They have skies and the skies have clouds, and some of them rain vanadium oxide and titanium oxide. They'll shine dimly from the small heat they have, maybe burning deuterium for a short time, and then when the gas runs out, they spend the rest of their lives alone and getting colder.

Meanwhile, Strauss and Fan's color-color diagrams kept turning up more quasars. They beat their previous record of redshift 5.0 with quasars at redshift 5.28, then 5.8, then 6.0, then 6.28, then 6.37. By 2001, they had 13,000 quasars; a year or so later, they found quasars at redshifts 6.1, 6.2, and 6.4; by 2005, they had more than 76,000 quasars, of which 19 had redshifts over 5.7, shining in their spooky light when the universe was a billion years old, 7 percent of its present age, just a toddler.

One reason for collecting more and more distant quasars was to answer astronomy's old question about the quasar cut-off, the time when looking back, we see no more quasars and conclude that's when they were first born. In the mid-1980s, Maarten Schmidt, Don Schneider, and Jim Gunn had found a cut-off around redshift 3.0 using 100 quasars. By 2006, Sloanies from several institutions who had been systematically surveying for quasars had selected 15,343 of them, lined them up chronologically, and found that their numbers increased with look-back time until somewhere between redshifts 2.0 and 3.0. Looking farther back, the numbers flattened out until, after redshift 5.0, they statistically died off. Schmidt, Schneider, and Gunn had found the flattening and a suggestion of the die-off, and Jim says though they were generally right, with only a few redshift 4.0 quasars, being right was a fluke. If the quasars had a let-there-be-light moment, it was somewhere beyond redshift 5.0.

Another reason for looking for the next most distant quasar was to find the Gunn-Peterson effect, the dark place in the quasars' spectra the young Jim Gunn noticed wasn't there, the sign that the first things hadn't yet lit up. The hot, early, ionized universe whose electrons and protons were running around unaffiliated was dense and opaque. After a few hundred thousand years of expanding and cooling, the universe's temperature declined enough for electrons and protons to stick together and make hydrogen atoms, and the universe became transparent to almost all light—except for ultraviolet light, which the hydrogen gas absorbed. Meanwhile, the hydrogen gas gradually collected itself into clouds and then into stars and quasars, which make their own ultraviolet light, which blew the electrons back off the hydrogen atoms, ionizing the universe again. But this time, the universe had expanded and rarefied and was so tenuous that it was transparent to all light—the ultraviolet, too—and the universe shone.

The quasar group, looking through their color-color diagrams for red things that might be quasars, either followed up the likeliest things on other telescopes, including the 3.5-meter at Apache Point, or asked colleagues with time on the large, sensitive, 10-meter Keck telescope to get spectra. The spectra of those quasars above redshift 6.0 were dark in just the right places, practically flat-lined, at the wavelengths at which ultraviolet light would be absorbed, to show the Gunn-Peterson effect, marking the time when the universe was still opaque but about to begin shining. Jim was pretty happy about finding this thing he'd predicted in 1965 when he was just a graduate student. He also thought they were just lucky that the cosmic dawn didn't happen at redshift 20, because the Sloan couldn't see much farther back than 6.0; even dawn at redshift 7.0 would have killed them.

In June 2001, data from the Sloan Digital Sky Survey was officially and publicly released, the Early Data Release—"Early" being an acknowledgment that though the data pipelines were still imperfect, the Sloanies needed to release the data anyway because they had promised the data would be available several years earlier. The publication had just under two hundred names on it: astronomers of course, some engineers, some instrument builders, some observers, a few professors, a lot of code writers.

By this time, the Sloan was doing enough science that the Sloanies stopped being itchy and frustrated and data starved, and everybody settled down. Peoples decided that the survey was finally stable, that the management organization he'd helped set up was working, that he'd done what he could do well, and that in the next few months, he would resign. Running the Sloan had been harder than simultaneously running Fermilab and closing down the SSC, and he was tired of dealing with Princeton. Intense as the conflicts had been, each one ended because the collaborators pulled back just in time to avoid a death spiral, and in 2002, Peoples retired content. "You know," he said, "the Sloan is a sort of a miracle. Most of these people did not have the slightest concept of how to work together. But everybody behaved well enough, long enough, that it succeeded." His only regret, he said, was that he was past the age where he could learn easily, so though he'd learned a lot about astronomy, he hadn't learned enough to be a real astronomer. He thought he'd like to try that someday.

Once Peoples retired, Rich Kron took over. Rich is modest, orderly, strict, and nonconfrontational. The collaborative atmosphere improved considerably and in fact, the project got a little boring. "Now we do all the stuff that projects are supposed to do," said Mike Turner. "We really got this down to industrial science."

Chapter 9

Precision

The data we got were *beautiful*.

—David Hogg, New York University

THE PIPELINE WARS had gone on for as long and as intensely as they had partly because those pipelines were unique. They effectively reduced a highly trained observational astronomer to code, and no one had written anything like them before. The Sloanies hadn't any notion of what they were up against until they started writing the code, and the subsequent tensions seemed more or less inevitable.

They succeeded nevertheless because the Sloan had an unusual convergence of astronomers whose code-writing talents were off-scale. These talents were scattered throughout the collaboration, though they tended to concentrate in the young. Astronomers had always written code, but these graduate students and postdocs seemed to the older astronomers to be some newly emerged population, specially evolved and born wired to write code. The code writers had grown up around DECs and Commodore PETs with

4k of memory and green phosphor screens; they spoke warmly of the TRS-80, the Trash-80. They'd tried to play computer games by moving little arrows around and got bored and decided to write their own programs. They'd read books—*Basic for Beginners*—and they could pick up computer languages faster than adults could, so they felt good about their privileged knowledge. When they got to high school or college, they'd taken a course or two and migrated off to learn different languages. With this education and these talents, some young astronomers chose, as their elders had, to concentrate on astronomy.

Others became astrocoders, a new and not-yet-adapted species of astronomer who writes code that takes and analyzes astronomical data, often for other astronomers to use. Sloanies were among the first cohort of astrocoders. Academic astronomers don't know what to do with them and treat them like soft-money engineers. Jim Gunn thinks that sooner or later, the community will realize that its lone-astronomer days are over, that the subject has become too vast and complicated to be done by anything except large groups whose science can be realized only through code.

One of the Sloan's young astrocoder/astronomers was David Hogg. In 1997, he'd been an Institute postdoc, suspecting that John Bahcall thought that his (Hogg's) astronomy was tired, and he wasn't sure he disagreed. Bahcall wanted him to get involved with the Sloan; Hogg didn't want to, but under pressure eventually went to some collaboration meetings and let David Schlegel talk him into worrying about the photometric pipeline. The worry was not about the pipeline itself, which by then Robert Lupton was writing and which was working beautifully, but about precision of the photometry going into it.

Photometry is measuring light. One of the first and most obvious things an astronomer wants to know about a star or galaxy is not how bright it actually is—that's complicated and involves knowing its distance—but how bright it appears to be. The Sloan measured brightness with an approach that had been used since astronomy began: choose one very bright star, traditionally Vega, to be the standard and measure everything else relative to it. So star X might be 0.0001 Vegas and galaxy Y, 2.5 billionths of a Vega. But Vega turned out to be much, much too bright for Jim's CCDs—"It doesn't actually burn a hole in the detectors," Schlegel said, "but you can't measure it." That's why the Sloan had the monitor telescope for calibrating brightness, and after the monitor telescope was taken out and shot, the Hopkins telescope had been installed as the photometric telescope.

Sloan's photometric calibration was persnickety and complex. The photometric telescope would measure the brightness of some galaxy that the 2.5-meter telescope would also look at, compare the star not to Vega (which was too bright even for the photometric telescope) but to an F subdwarf star named BD+17, which had already been calibrated to Vega, then measure everything else that affected the galaxy's brightness—atmosphere, dust, pollen, humidity—then adjust the galaxy's brightness accordingly. Then it would double-check its adjusted brightness with a similar measurement done at the U.S. Naval Observatory's telescope in Flagstaff. Finally it would, in effect, tell the photometric pipeline that star X was really 0.00014 Vegas. Calibrating this way, Sloanies estimated that they were getting the brightness of stars right to 2 percent. Michael Strauss's science requirements had specified 2 percent too, so most Sloanies were happy.

Schlegel, however, thought this whole system of calibration was ridiculous on the face of it—too many steps, each of which introduced its own errors. And Jim was worried about the extent to which the differences between those three telescopes—the 2.5-meter and the photometric telescope and the Navy's instrument at Flagstaff—were affecting the photometric precision. So on July 12, 2000, Jim wrote an e-mail to the collaboration members that began, "It is the opinion of several of us that the photometric calibration is the single tallest pole in the way of our reaching full operation."

The e-mail announced the solution: David Hogg at the Institute for Advanced Study would lead a group "to address these problems from a completely independent point of view." The group would be called the photometric task force, and would consist of Hogg, David Schlegel, and Doug Finkbeiner. The team would investigate the whole photometric process, find the problems, and write independent code to fix them, the e-mail said; and they'll start with a phonecon that afternoon. "Thus saith the Lord," said Finkbeiner.

Hogg's presence on the team could be seen as the Institute's contribution to the collaboration, and happily, he hadn't antagonized anyone yet. The photometric task force's assignment was to go to Apache Point and, as Hogg thought of it, "figure out what's going on that we're not doing as well as Jim Gunn thinks we should be doing." The software they then wrote—which Schlegel named hoggPT, "PT" for "photometric telescope"—eventually cut out the need for the Flagstaff telescope. "It was a slick little package," said Finkbeiner. They finished it by the end of 2000 and everyone liked it.

By running hoggPT they had learned exactly what precision the photometric telescope plus the site on Apache Point were capable of—not the 2 percent of the science requirements but 1 percent. Before Sloan, expert photometricians could get 1 percent precision on a few objects in half a night, but no one believed such precision could be achieved by a survey. After running hoggPT,

Sloanies understood that nothing was in the way of 1 percent precision except good software. So instead of using the photometric telescope, which was not as good as the 2.5-meter with Jim's camera, they could write software that would let the 2.5-meter calibrate its own excellent self.

Finkbeiner and Schlegel knew of a technique used by astronomers looking in the cosmic microwave background for 1-in-100,000 variations, and thought they could apply the same technique to the Sloan. The stripes of the sky drift scanned by the 2.5-meter are taken on different nights. And from night to night, the camera and telescope electronics change minutely, the mirror collects a little more dust, one CCD might have gotten a thousandth of a degree warmer—the upshot being that a twelfth-magnitude star in one stripe looks slightly brighter or dimmer than a twelfth-magnitude star on another stripe, that is, the stripes have different photometric zero points. The idea was to get the whole sky and everything in it to share the same zero point.

So, Finkbeiner and Schlegel thought, direct the telescope to scan perpendicular to those stripes, and where the stripes and perpendicular scans intersect, find the difference between the zero points. Then, using mathematical matrices and linear algebra they could calculate a universal zero point, in effect, resetting all Sloan images to all other Sloan images. Then they could cut out the photometric telescope and tie the 2.5-meter's entire collection of images to BD+17.

The perpendicular scans could be taken quickly, zipping the telescope up and down the sky at seven times its normal speed. Because in the pattern of scans and stripes Polaris looked like the hub of a wheel, the scans looked like spokes, and the stripe at the horizon looked like a rim, they called their idea the Apache Wheel. The code that controlled the whole operation they called *überkalibration*—"Finkbeiner and Schlegel are good German names," said Schlegel—then anglicized it to ubercalibration, and later shortened it to ubercal. Eventually they worked with other astrocoders, notably a Princeton graduate student named Nikhil Padmanabhan. In tests of ubercal, a 12.33 magnitude star in one stripe is 12.33 magnitude on another stripe to within 1 percent.

Sloan's first stated mission had been to find the universe's large-scale structure, the way galaxies lived in clusters and clusters collected into superclusters, and how big the superclusters were. At the time, observers were mapping individual superclusters by mapping galaxies, but mapping galaxies one by one wasn't covering much of the sky. To cover more sky, they wanted to map the middle child of the large-scale structure, the clusters of galaxies. So back in the mid-1990s, the first thing that the Sloanies in the Cluster Working Group had to do was figure out how to find clusters, specifically, how to write the code so that a pipeline could spit out clusters. This wasn't going to be easy—how does a pipeline identify a bound group of galaxies?—and they couldn't agree how to do it best.

Maybe instead they could identify clusters by finding the brightest cluster galaxies, BCGs, the same unpredictable galaxies that Jim had once hoped would be a good standard candle. Brightest cluster galaxies are abnormal; they're a kind of central garbage pile, found near the gravitational centers of clusters whose galaxies have been colliding, merging, fragmenting, falling inward. But how to find them? They were known to be unusually large, always elliptical, and always red—that is, no longer forming new blue stars. Fermilab's Jim Annis, who had done his PhD on clusters, said that the survey should simply look for the biggest, brightest, reddest galaxies—well within the capabilities of a pipeline—and he could almost guarantee a cluster around each one.

Once you've found the biggest, brightest, reddest galaxies, you've found the clusters, and the clusters will align along the superclusters. So obviously, the brightest, reddest galaxies are tracer particles for the superclusters that outline the large-scale structure: map those galaxies, and you'll trace large-scale structure. For a while, the Cluster Working Group called these galaxies BRGs, which stood either for big red galaxies or, preferably, for bright red galaxies. Daniel Eisenstein, who'd been a Princeton graduate and was now an Institute postdoc, suggested instead they be called LRGs, for luminous red galaxies because astronomers talking among themselves use "bright" to mean how much light we see and "luminous" to mean how much light the thing is giving off. The LRGs, though they were intrinsically luminous, were so far away they weren't very bright.

Eventually the Sloan used its LRGs-as-tracers idea, mapped the large-scale structure, and found what the English-Australian 2dF survey had already found. The Sloan and 2dF maps showed superclusters, not isolated in clumps but parts of a universal network, filaments of lights that are denser or thinner and sprawl over sheets that fold themselves around dark voids. Large-scale structure looks like a foam of lights and black bubbles that expands to fill the universe. It looks like solidified lava, or a sponge, or medically imaged tissue, or a night flight's view of cities and streets and suburbs. It's almost a shock: it looks normal. It's biological, geological, natural, just the way you'd expect the universe to look.

Sloan and 2dF together finished off the large-scale structure problem: the average size of a filament or void is around 300 million light-years. And in fact, by the time the Sloan published its findings, the structure's size was no longer much of a mystery.

But the universe's large-scale structure wasn't quite that simple. Since 1970, theorists had been predicting another, subtler structure, superimposed on the 300-million-light-year structure and called baryon acoustic oscillations. According to the well worked out theory of the Big Bang, all structure was born inauspiciously, in the early universe, in a shapeless and violently hot plasma made of radiation and all the particles of matter, some of which was the ordinary matter of stars and people. The plasma boiled; radiation and matter piled up into little drifts, blew back out, piled up again, forming and reforming drifts and ripples of radiation and matter.

But as the infant universe expanded the radiation cooled, until finally it separated from matter and went its own way, the last pattern of little ripples still impressed into it. The universe kept expanding, and radiation kept cooling, its wavelength getting longer and longer until it reached the long wavelengths of microwaves. And its pattern of ripples, in a map made 13 billion years later by a satellite called the Wilkinson Microwave Anisotropy Probe (WMAP), looked like hot and cold patches in the cosmic microwave background, accurate to an astonishing 1 percent.

That last pattern of little ripples was not only in the radiation, it was also in the ordinary matter. And after the radiation decoupled from the matter, the matter ripples had been free to feel gravity; they collapsed in on themselves, attracted more matter, grew and turned themselves into galaxies. Because the radiation and matter had rippled together, the pattern in the microwaves should exactly match the pattern in the galaxies. The physics of gravity and expanding waves are well known, so once WMAP had measured the microwave background, theorists could safely predict that the ripples in microwaves that were 1 degree across should correspond to ripples outlined by galaxies that are 450 million light-years across.

Those particular ripples—other ripples in matter and radiation behaved differently—behaved like sound waves, so theorists called them acoustic oscillations. And because physicists' name for the ordinary matter in galaxies is baryons, the ripples' formal name was baryon acoustic oscillations, BAOs for short. If you could hear them, they'd sound like static.

By 2004, the Sloan had found the redshifts of 46,748 LRGs scattered over 3,816 square degrees of sky back to a redshift of 0.47, and then they looked for the patterns they made. You can't look at the Sloan map and see baryon acoustic oscillations outlined by LRGs. In the first place, as Daniel Eisenstein explained, the ripples don't look like those from a stone thrown into a pond, they look like thrown gravel—a mess of interacting, overlapping ripples. In the second place, the baryon ripples are shallow and show up only statistically, only in the math. Nevertheless, in 2004, the Sloanies found them there, 450 million light-years across.

But like Hercules and the Hydra, every time astronomers smack off one of the monster's heads, the monster just grows more. In this case, the next Hydra head is much, much worse. During those years that the Sloanies were thinking about clusters and writing ubercal, the measurement of the universe's expansion had gone from a well-defined problem with an unequivocal answer to a piece of nastiness undermining both cosmology and its underlying physics.

The Big Bang theory clearly laid out the broad outline of the universe's history. The universe was born explosively, the explosion expanded and cooled until the matter in it could begin gravitating, and ever since, under the mutual gravitational attraction of all the universe's matter, the expansion had been gradually slowing. But in 1998, two separate and competing teams had found that the whole notion of gently slowing expansion was true only in the distant past. Beginning cosmologically recently, somewhere around redshift 0.5, about 5 billion years ago, the universe stopped slowing and began expanding faster with time; it was accelerating.

The obvious question was why. What was making the expansion accelerate? Was something pushing it? The cause of the push, Michael Turner—who is part poet—called "dark energy." But that was just a name; what dark energy actually was, nobody had or has a clue.

And because cosmologists' understanding of the universe is based on physics, the cluelessness extends to physics, mankind's most basic science. Not that the physicists don't have theories anyway: Maybe it's just some sort of energy they hadn't known about. Maybe gravity doesn't work the way Albert Einstein's theory of general relativity says it does. Maybe Einstein's famous cosmological constant—which he put into the equations of general relativity to fudge the universe's expansion and which he later took back out—was right after all and space/time has some weird springiness that acts like antigravity, and we call it dark energy. Maybe if physicists are ever able to understand the quantum physics of the exceedingly small in the same terms as the gravitational physics of the exceedingly large, they'll come up with some other reason that the expansion of the universe is accelerating. Jim says that dark energy has one theory for every physicist who thinks about the problem, and he suspects that all the theories are at some level untenable. Dark energy means something is wrong with physicists' and cosmologists' understanding of the universe, and whatever it is, it's basic, it's immense, and it makes science fiction look mundane.

Each of physicists' theories makes different predictions about the acceleration. So the first step in figuring out dark energy is to measure the acceleration to within an inch of its life. The original 1998 measurements of acceleration were done with standard candles that are supernovae of a kind called Ia, stars that die by exploding with a certain, known brightness. Between 1998 and 2003, astronomers measured and remeasured the universe's acceleration by the same and different techniques—prove the acceleration wrong and you're famous. Most of their measurements were of Ia supernovae, at first of those nearby, around a redshift of 0.05. Later surveys were farther away, out between redshift 0.5 and redshift 1.0. Both measurements, near and far, had problems: the nearby Ia's were in a small look-around volume in small numbers with small errors; the distant ones in a large volume had large numbers and large errors. To seriously constrain the acceleration, you needed Ia's in the middle redshifts in huge numbers with tiny errors. The Sloan hadn't been intended to survey supernovae— it had found them mostly because they were in the images along with the galaxies. But once the Sloanies understood that measuring the acceleration depended on finding supernovae, they began collecting them in earnest.

The usefulness of Ia's as a standard candle depends on the way the explosion first brightens and then dims, so using Ia's to measure acceleration means revisiting them time and time again. For three months at a time over a period of three years, on as many nights as possible, the Apache Point telescope scanned the same stripe, stripe 82, on the sky. It found five hundred Ia's—about the same number as all the other supernova surveys put together— in the middle redshifts between 0.1 and 0.4. The analyses of the Sloan Ia supernovae also found the universe accelerating, just as the earlier studies had, but the acceleration was measured now to the Sloan's photometric precision.

Once the Sloanies found the baryon acoustic oscillations, however, they understood they had another way of measuring acceleration, this one independent of the Ia's. In this case, the standard isn't a candle, it isn't brightness; it's a ruler, it's a length, those

450-million-light-year ripples that are baryon acoustic oscillations. Farther back, at a redshift of 0.6 say, the ripples would be smaller and the ruler would be shorter; farther back yet, say at redshift 2.5, the ripples would be smaller and the ruler shorter yet. So around 2005, the Sloanies began planning to extend their survey to 1.5 million LRGs at those higher redshifts, watch exactly how the ruler shortens, and find out how the acceleration seems to behave and whether dark energy has changed with time.

They call their extension the Baryon Oscillation Spectroscopic Survey, or BOSS. BOSS will measure acceleration independently, with the enormous numbers and the remarkable precision that only the Sloan has.

About half the science the Sloan does is science it had promised in its first proposals; the other half is science no one had thought of. At the beginning of the project, no one had known the universe was accelerating, no one had heard of dark energy, and baryon acoustic oscillations lived in the small details of theory; certainly no one had measured them. Any survey will find the unexpected—"You write such things in proposals," said Michael Strauss, "but we did not imagine the extent to which all the things we hadn't imagined would be possible." And a lot of that unexpected science wouldn't have been credible, let alone even discoverable, without the Sloan's homogeneity and precision. If anyone needed an excuse for unyielding do-it-rightness, for writing e-mails in capital letters, for side-stepping management, for second-guessing your colleagues, or for polishing brass knobs, it's a survey so catholic, systematic, and beautifully calibrated that its measurements are statistically certain and precise to 1 percent. And how did it happen then that with a firm agreed-upon, systems-engineered requirement of 2 percent, they got 1 percent?

Because, Bill Boroski the new project manager smiled and said, "you can't stop these guys."

And in a turn of events out of a Dickens novel, the person in charge of Sloan's new BOSS project—of making work breakdown structures and writing science requirements and creating schedules—is David Schlegel. Now he writes polite e-mails like: "Dear Martin, Here are my suggestions for revising the BOSS section of the agency proposal. As usual, you have written a lucid first draft. I hope you will be able to get a revised version back to me by Friday; sorry I couldn't get feedback to you earlier. You have 1–2 pages to expand if needed. Cheers, David." Schlegel thinks of his new role in life as a director as his punishment, as justice. "It's not fun," he says. But he thinks the atmosphere on the project is collegial: "Here everyone's on the same team," he says, "making sure we get the right things done."

BOSS relies mostly on spectra, and though Schlegel was partly responsible for Sloan's spectro 1D and 2D pipelines, the spectra that came out of them didn't have the precision that the photo pipeline/ubercal did, the 1 percent. The Sloan spectra were good only to 2 percent. So Schlegel's still declaring that the spectra aren't as good as they should be, still saying, "It has to be done and we're the people to do it," only now he's got a team, one of whom calls this phenomenon "spectro-perfectionism." Schlegel announces to his BOSS team, "We're gonna do way better than two percent."

Chapter 10

The Virtual Observatory

The whole game is like almost a religious undertaking that people will devote their lives, their time, their wits for things which have no practical importance. And there's something rather beautiful about that. Look, we think of it as a big planet but it's really just a piece of dust out there. And on this piece of dust, these creatures are walking around, 5 billion of them. And 5 billion minus 5 thousand are looking down at the earth, and 5 thousand are looking up at the whole rest of the universe. So there's a lot to look at per square astronomer.

—interview, Jerry Ostriker, Princeton University, 2007

DATA ARE TO ASTRONOMERS what money is to the rest of us and accordingly, astronomy has always been divided into haves and have-nots. When the Sloan started in the late 1980s, the country had around 4,500 astronomers and around ten good-sized telescopes, the large majority of which were, like Caltech's Palomar Observatory, private. Astronomers at universities with private telescopes went to the mountain, observed, and returned with their plates, which they put into their cupboards and analyzed at their leisure. No one else saw the plates. The "great professors"—Jerry Ostriker's term—staked out their claims; they were, as one observer said, the eyes on the heavens for the entire human race.

The alternative for the 60 percent of astronomers without private telescopes were NSF's two national observatories, whose time was oversubscribed by four or five to one and which awarded any given astronomer two to ten nights a year; if the nights were cloudy, apply again next year. Those astronomers also take their data home with them, but because the observatories are supported by the taxpayers, the time astronomers have to analyze data themselves is limited to a year, maybe eighteen months. After that, the data has to go into an archive anyone can use. But in the 1990s, even at the national observatories, archiving seemed impossible. As Jim explained, "the data are worthless without proper knowledge about exactly how to calibrate them and exactly under what conditions they were taken and exactly what the problems with the instrument were at the time."

The exception was NASA, with its publicly funded, well-calibrated archives of data from its infrared, X-ray, and ultraviolet telescopes in space. The national optical observatories did eventually set up an archive; the country was about to build some new, big public telescopes; and universities were increasingly buying their own telescopes by joining together into consortia, such as ARC. Eventually the aristocracy would have been infiltrated, even without the Sloan. But the Sloan archive was about to set off a democratic revolution—or it would have, if anyone had known how to build a searchable archive housing unprecedented amounts of data.

Throughout the 1990s Sloanies had enough to do just to get the data. The archive could come later, when the software was done, when they had some time. But when they did finally finish the archive, the Sloanies thought, maybe they could put it on compact discs, which they could sell, maybe for $20,000 a set. That strategy certainly worked for the Palomar plates, but you might suspect that cosmologists had not yet quite wrapped their minds around the cheapness, portability, and universal accessibility of digital information.

When Fermilab signed on in 1990 and agreed to write the software, it seemed to be the natural home for the archive, though Fermilab demurred from funding such a thing, and the survey had no budget planned for it. In 1992, Don York wrote to Jerry Ostriker that "maintenance of the archive as a living archive has not been addressed or funded, and any software needed to do anything complicated must come separately." But then Hopkins had joined, and along with responsibility for the spectrographs came hints and murmurs that it was also interested in putting together the archive. Hopkins's Alex Szalay liked statistics and computers.

The issues of hosting and funding were apparently not burning, because by 1994, they still weren't settled. Meanwhile, Alex and his postdoc Andy Connolly had calculated that the data including images were going to add up to around 12 terabytes—"tera-" is "trillion"—and somehow, some astronomer should be able to ask the archive something like "Find all blue galaxies near quasars." Connolly wrote to the collaboration members that the database at the Space Telescope Science Institute was only 100 gigabytes— "giga-" means "billion"—and a search took five hours. "Unless we have an efficient means of searching the data," he wrote, "our ability to do the science we want will be severely compromised."

Toward that end, they'd been looking at commercially available software—like Versant, ObjectStore, Objectivity, Oracle—for organizing and searching databases. In the mid-1990s Sloanies didn't know much about databases, but ignorance didn't keep them from having opinions and arguments. They argued over which of these new start-ups was most likely to go belly-up and how much they'd be at the mercy of the database's company changing the version on them. Steve Kent liked Versant, Alex liked Objectivity. So Fermilab and Hopkins, with help from Princeton and the Naval Observatory, compared them, and Objectivity won. Nobody was particularly happy about it. And they still didn't know how to make the 12-terabyte-archive public—the NSF was nagging them about it—other than to record it on stacks of compact disks at, say, 600 megabytes each, or around 19,000 CDs altogether.

Alex Szalay had come to Hopkins from Hungary in 1987, the same year Jim got the idea for the camera. His first name is really Sandor; Hungarians call him Sanyi, pronounced "Shani," but the rest of the world calls him Alex. He has a long, flat, alert face and talks with almost no inflection. He'd seem impassive if he ever sat still. He's nervy and intense; when he gives a talk, he's all over the stage, which makes a kind of sense because before he left Hungary he was a member of a jazz/rock band called Panta Rhei that performed, made recordings, and sold lots of them. Szalay began his scientific life as a mathematically inclined cosmological theorist— Budapest was a central meeting place for theorists from the Soviet Union and the West—and had been a postdoc at both Berkeley and Chicago. When cosmological theory began hitting the wall for lack of cosmological observations, Alex turned observer and worked with Rich Kron finding large-scale structure with pencil beams. He joined the Sloan, he said, because "at the age of forty— actually forty-three—one would like to do things which have, I don't know, some kind of legacy."

Back in Hungary, Alex's parents, both physicists, were good

friends with another physicist named Charles Simonyi, who had a son also named Charles around Alex's age, whom Alex never got to know because the younger Charles had left Hungary for good at age sixteen. Young Simonyi ended up at Stanford, then moved to Microsoft, where he led the development of Microsoft Office, with its juggernauts Word and Excel. Microsoft Research has its headquarters outside Seattle, so when the 1997 Sloan collaboration meeting was scheduled for Seattle, Alex's and Charles's parents decided their sons should finally meet. Over dinner, Alex told Simonyi about the Sloan and its database problems, and Simonyi told Alex about his friend at Microsoft Research, Jim Gray, who was good at databases—in fact, Simonyi called Gray up then and there. A few weeks later, Alex met Gray.

Jim Gray had worked all over the information technology industry, had won awards and prizes, and was now building a website called TerraServer, which put together images from the U.S. Geological Survey satellites into an archive in which you could search for pictures of the Vatican, Alcatraz, or your own driveway. TerraServer could be seen as an analog to the Sloan archive— searchable objects mapped on a sphere, only with TerraServer you were looking down on the sphere, and with the Sloan, up. Alex thought of the Sloan archive as the TerraServer turned inside out. Gray would work with the Sloanies and wouldn't charge—his job at Microsoft allowed him to do what he wanted. Gray liked the Sloan because its data were more interesting than Walmart's.

Gray thought, as did Alex by this time, that Objectivity wasn't going to work—too many features that Sloan needed were missing—and he suggested that the Microsoft SQL Server might be better. So Gray and Alex, over the Christmas holiday, wrote code that would convert Objectivity to the SQL Server, added enough astronomy that they essentially had an astronomy library running inside the database, and in the process, created a query system unique to the Sloan. The collaboration, along with much of the scientific community, disapproved of Microsoft's reputation for mandatory, unnecessary, and buggy bells and whistles and took several years to officially convince itself to use SQL—Structured Query Language—for the archive.

In 2000, Intel gave Alex a grant to buy the computers to hold the archive, and they wanted him to give a show-and-tell using the data. Jim Gray advised Alex that best show would not be to type a query into a database and get back a row of numbers, but instead to make a pretty website. Alex did a first design of the website and showed it to his thirteen-year-old son, Tamas, who said that no decent young person would consider it and that Alex should try for something that looked like a gaming interface. Alex and Gray thought that if the Sloan didn't want their SQL archive, maybe kids—and by extension teachers and school systems—would. So Alex and Gray put together a prototype of what became the Sky-Server, Sloan's public presence in the world.

In June 2001, after ten sleep-deprived days, Alex and the post-docs got the first year's data into the SkyServer, and the survey was released publicly for the first time, the Early Data Release. The Early Data Release was 80 gigabytes of data, with 14 million objects and 50,000 spectra. They hadn't needed the 19,000 compact disks after all. Jim Gunn was pleased at the prospect of astronomy done from an archive. "People dreamed of doing it this way," he said, "but had never been able to do it." Between June and October of 2001, the SkyServer website had 1.5 million hits; the number of astronomers in the country was now around 6,500.

Alex and an astrocoder at Hopkins named Ani Thakar continued refining SkyServer, Alex working eight-hour days, Ani twelve. Jim Gray and his colleague, Microsoft's Curtis Wong, continued working with them, and Alex watched the SkyServer come to life. Gray thought of his work for the Sloan as a long field trip. The survey's data were not only interesting, he said, but capitalistically worthless, free and completely without commercial potential. "People at Microsoft think I'm a communist," he said. Gray wanted SkyServer to be an archetype for dealing with the enormous amounts of data coming out of the likes of computational biology, computational physics, or computational neuroscience. He saw SkyServer as an example of a new and democratic way of doing science that he called eScience.

Anything you want to know is in the SkyServer, everything that's in the surveyed sky, every image in five colors, everything that a spectrum can tell you, every asteroid, comet, star of any kind, galaxy, and quasar, sorted by color or brightness or distance or age or composition or neighborhood.

Start by finding the latest version of the SkyServer: though the latest version includes all previous data releases, the Sloanies don't throw the old versions away. Click on Search, then Radial, meaning search within a certain radius. Specify the part of the sky you want to see as though you're pointing a telescope, by using astronomical coordinates—right ascension and declination, RA and dec—and a radius in arc minutes, hit Submit. You get back a table with a handful of objects coded by number—6's are stars, 3's are galaxies—and their brightnesses in each of Jim's five color filters, along with the errors in each brightness. Galaxy SDSS J130200.57+023013.8 in the ultraviolet filter is magnitude 23, error is 0.7 magnitude either way, but that u filter always had more problems than the others. The same galaxy in the green filter has a magnitude of 22.54, error is 0.18 magnitude. Click on SDSS J130200.57+023013.8 and you get a picture of it sitting in crosshairs, which you'll need because it's all but invisible. If its spectrum was taken, you can click on that too. Click on another entry in the table, SDSS J130200.81+022942.6, and it turns out to be a star, a poufy red ball in the crosshairs. All these objects in this table are neighbors in the sky, and you might have found interacting galaxies, or binary stars, or a cluster, or the host galaxies of quasars.

Then you can do something a telescope can't do. Back out on the SkyServer home page, click on Sample SQL Queries—assuming you can't write your own SQL query—and pick, say, all the extremely red galaxies or the white dwarf stars within a 0.5 arc-minute radius, and copy and cut the commands. Go back to the home page, click on Search, SQL, then paste the commands, get a list of all extremely red galaxies or all white dwarfs in a patch of sky. Get their spectra, if they were taken, and download the whole shebang to your own computer. Or go find asteroids by their names—Ceres, Pallas, Juno, Vesta—and where they are in the sky and then map them compared to the planets and see the whole solar system as it appears on any given date. Or find galaxies that you classify by shape, then locate on the sky, then find their distances so you see which ones live in the same cluster, and discover for yourself that cluster galaxies are ellipticals.

After the Early Data Release, the subsequent releases came more or less regularly every year. The Sloanies, as was their culture's custom, meant to keep each batch of data proprietary, for private use, for a year before turning it over to the community. But they kept running late, and they'd made promises about public availability to the NSF and the community, and the proprietary year shrank to six months, then two or three.

> Data Release 1, June 11, 2003: 53 million objects and 186,250 spectra.
>
> Data Release 2, March 15, 2004: 88 million objects and 367,360 spectra.

Data Release 3, September 27, 2004: 141 million objects and 528,640 spectra.

Data Release 4, June 27, 2005: 180 million objects and 849,920 spectra.

Data Release 5, June 28, 2006: 215 million objects and 1,048,960 spectra.

Data Release 6, June 29, 2007: 287 million objects and 1,271,680 spectra.

The NSF's program officer in charge of astronomy, Wayne Van Citters, watched the images and spectra flooding out and thought that the Sloan must be making life fairly depressing for the astronomer with several years on a 4-meter telescope getting tens of spectra on one class of stars. It reminded him of a Peanuts cartoon about show-and-tell day at kindergarten, when Linus shows his copies of the Dead Sea scrolls and explains, "It's a portion of I Samuel 23:9–16. I'll try to explain to the class how these manuscripts have influenced modern scholarship." Then it's Charlie Brown's turn, and he just says, "Well, I had a little red fire engine here but I think I'll just forget it."

Data Release 7, October 31, 2008: 357 million objects and 1,640,960 spectra which are classified into 929,555 galaxies, 121,373 quasars, 464,261 stars. Those 464,261 stars were there in the first place mostly because they wouldn't get out of the way: they're in the galaxy, the Milky Way, in which we live, and seeing any other galaxies means we have to look through our own stars. So the Sloan got a lot of data on the Milky Way.

The Milky Way had been at the top of no one's list of the parts of the universe to be mapping. A 1997 list of key projects for the Sloan doesn't mention the Milky Way, though an early version of the Principles of Operation suggested that studies of the Milky Way might be considered if they didn't interfere. The Milky Way was known to be a spiral galaxy with a bright core and pinwheel arms forming a disk full of newly born stars. The core and disk were, and still are, hard to study because they're full of dust that gets in the way of seeing the stars there. The disk was known to be surrounded by a dim, sparse halo of old stars and odd little balls of old stars called globular clusters. The halo was generally a mystery: no one was sure where it came from or exactly how it was shaped. Astronomers had collected around one hundred halo stars, either nearby or in pencil beams. In the early 1990s the Milky Way was a moderately complex topic in standard textbook astronomy.

Of the people interested in the Milky Way, only some were interested in the galaxy itself. Others were interested more generally in how galaxies form, thinking that maybe the Milky Way held the history of its own formation. The study of galaxy formation in the early 1990s was a mess, more concept than science. You could try to compare nearby old galaxies with distant young ones to see any differences, but the distant young ones were too dim and rare to do much good. You could try to find baby galaxies just being born, but even when you got a spectrum of something at high redshift with brand-new stars, you couldn't see it well because the most telling part of the spectrum had redshifted right out of the optical into the infrared, where the sky was bright and the CCDs insensitive anyway. You could try to theorize about newly forming galaxies, put a moving gas cloud and gravity into a computer simulation and watch it spin up and take shape and start making stars, but for realism you had to add radiation and feedback and magnetic fields and do it all with equations on computers not powerful enough to handle it all. The Milky Way might have held the closest evidence of galaxy formation, but unlike any other galaxy you could look at, it was spread across the sky, and you needed a survey to see it.

About the time the Sloan started, people were beginning to suspect that galaxy formation, however it happened, might have a lot to do with mergers. Maybe galaxies started in crowded neighborhoods and littler ones were attracted to bigger ones and fell into them, making the bigger ones bigger. Observations did turn up galaxies that were obviously under each other's influence, the spiral arms of one distorted and reaching for the other. Theory predicted that galaxies like the Milky Way would be surrounded by nearby satellites—dim, pathetic galaxies called dwarfs, waiting to merge. Sloanies—notably those at Cambridge University in England, which had recently joined the Sloan—looked for dwarf galaxies and eventually found fourteen of them, more than doubling the number of known ones. "They should have been there, nobody had been able to find them before," said Jim happily. "We're finding them in droves." They're named for the constellations in which they occur: Canes Venatici, Boötes, Leo, Virgo, Coma Berenices, Ursa Major, and Hercules. One dwarf galaxy, the first one Sloanies found, was so faint it may have been an out-of-place globular cluster; it was named Willman I for the graduate student, Beth Willman, who found it. Another, Leo V, was right next to Leo IV but hadn't been seen before because it happened to be in the line of sight to a distant cluster of galaxies.

So maybe galaxies did form by mergers, and if so, then evidence of mergers might show in the Milky Way's halo. The halo had been nicely represented in the survey: the camera had avoided the disk as being too crowded to differentiate galaxies, quasars, and stars, but it had covered most of the halo in the northern sky, and it had recorded a swath of the halo 2.5 degrees wide and 100 degrees long in the southern sky. It collected 5 million stars in five colors each. Heidi Newberg and Fermilab's Brian Yanny, thinking they were going to map a smooth, uniform halo, started counting stars—blue ones, bright and dim—and recording their locations. In 2000 and 2002, they found the halo was less a halo than a clumpy congregation. "We took a long time to see the halo wasn't smooth," said Newberg, "then we took a long time to believe it."

In particular, running through the halo were two streams of stars of the same color. Those star streams were also the same color as a small, undistinguished galaxy called the Sagittarius dwarf in the Milky Way's neighborhood. The Sagittarius stream had been seen sketchily a few years earlier and now was followed up by postdocs at Cambridge University and the Max Planck Institute in Astronomy in Heidelberg. Sloanies and some non-Sloanies using the data releases joined in looking for more streams and found them. The streams the Sloanies saw were apparently the result of the Sagittarius and several other dwarf galaxies falling into the Milky Way, and in the process coming apart into so-called tidal tails that wrapped all the way around the Milky Way.

Depending on how "streams" are defined, the Milky Way has at least a half dozen to a dozen of them, some named for the constellations through which they seem to run (Sagittarius, Monoceros, Cetus, Virgo), some for rivers of the underworld (Styx, Acheron, Lethe, Cocytos), and one called Orphan. Sloanies called the map of the halo the Field of Streams. The Sloanies from Heidelberg who looked at the halo's globular clusters—dense, tightly bound perfect balls of stars—saw long streams coming out of them too, pulled out by the Milky Way. The streams coming out of dwarf galaxies seem to be broader than the streams coming out of globular clusters. One globular cluster, called Pal (for Palomar) 5—the one Connie Rockosi wanted for her thesis but didn't have spectrographs for—is likely to have lost all its stars and be gone in 100 million years.

The Milky Way's halo looks less like a sphere than it does like arching banners streaming over and around the spiraling disk; it looks like it belongs in a parade. Each stream is made of stars of differing ages and chemical compositions, and the stars moving in each move at different speeds. The implication is that each stream had a different source, its own broken dwarf galaxy, now merging into the Milky Way. The dwarfs would be gone except that their stars remember them, running together in colored, co-moving streams.

Newberg thinks that because the Milky Way survey was part of the Sloan, part of a study of a much larger part of the universe, cosmologists noticed it quickly and then paid new attention to it. Clearly, if the Milky Way formed by merging with lesser entities, the rest of the universe's galaxies could have too. Constraining grand cosmological questions with local observations is now given the paradoxical name, near-field cosmology.

In 2003, in March alone, the Sloan archive had a million hits; the total number by that time was 10,682,584. By 2006, it had 217 million hits. The number seemed to be doubling every year. By 2010, it was 713,581,441. It had 60,000 to 70,000 different users every month—the number has risen steadily since 2001—and the country now had maybe 7,000 astronomers. The next most interesting question, after what science is in the archive, is who besides the Sloanies is using it.

One group was an army of non-Sloan astronomers. Two-thirds of the papers using Sloan data have been written by non-Sloanies. A paper using the first data release found that a merging pair of dwarf galaxies, called Arp 305, had forty-five clumps of stars that were just forming but only a few clumps of stars that were older, suggesting that merging dwarf galaxies didn't last long. One of the paper's authors was from East Tennessee State University, another was from Iowa State University, and neither institution is nominally an astronomical powerhouse. Between 2000 and 2009, of all the astronomers in the world, one in four or five has used Sloan data. Sloanies noted that the archive's usefulness is limited not by the survey's original intent but only by its users' inventiveness.

Another is a brand-new group, the virtual astronomers. Back in the mid-1990s, before Alex Szalay and Jim Gray met, Alex had been talking to astronomers at an all-sky infrared survey called 2MASS about Sloan and 2MASS using the same structures for their archives so that the sky could be accessible in both wavelengths. Maybe, Alex thought, they could merge the Sloan survey with several other surveys in still other wavelengths to make a digital map of the sky in all wavelengths.

Astronomers have always segregated themselves by wavelength: infrared astronomers rarely talked to ultraviolet astronomers, nor did they use each other's data. The reason was not narrow-mindedness, but incompatible technology: the length of the waves in the whole electromagnetic spectrum of light range from gamma rays that are the size of the nucleus of an atom, a quadrillionth of a meter, to radio waves measured in meters; as a result, the telescopes for each band of wavelengths must be wildly different. Radio astronomy is often done by coordinating giant dishes spaced out over miles of ground; X-ray astronomy is done by levered cylinders in space. Even the units in which the light is measured are different: radio is measured in Hertz, infrared in microns, optical in angstroms, X-ray in kiloelectron volts. The wavelength communities have different technologies, different cultures, and hardly speak a common language. But what goes on in the sky goes on in all wavelengths, with different wavelengths emerging from different processes: radio waves, for instance, can come from strong magnetic fields, infrared from dust at cool temperatures, ultraviolet from starbirth, and they can all be coming at the same time from the same object. So by looking only in one wavelength, astronomers were limiting themselves, were listening to a symphony and hearing only the violins.

While the Sloan was getting itself started, a slew of surveys were also beginning or running, some like 2dF in the optical but most in other wavelengths. FIRST surveyed the sky in radio wavelengths, 2MASS in infrared wavelengths. GALEX in ultraviolet. ROSAT in X-rays. All these surveys had archived their data and, for the most part, made their archives public. So what Szalay and Gray were trying to figure out was how they could put all that data in all those wavelengths into one archive and hear the whole orchestra.

Other astronomers had also been thinking about multiwavelength sky surveys. The technologies for sensing, computing, storing, and electronically sharing data had all been evolving on fast forward, and the surveys were producing huge amounts of data. Rather than going from archive to archive, downloading a bit of X-ray data, a bit of optical data, and putting it together in their own computers, astronomers needed to get all the data they might want from one place. Astronomers at Caltech were calling it the Digital Sky. By 2001, seventeen institutions had collected themselves into a National Virtual Observatory. Meanwhile the Europeans had been doing the same, calling theirs the Astrophysical Virtual Observatory, and in 2002 the virtual observatories merged into the International Virtual Observatory Alliance. "Everyone, their grandmothers, and their kitchen sink is in the virtual observatory," said Alex. "Everyone in this country and in the world."

Željko Ivezić, who eventually got a faculty job at UW, was interested in stars that shine most brightly in the optical but are often surrounded by dust, which radiates in the infrared—"dusty stars," he calls them. Before the multiwavelength archives, to see what his optical stars looked like in infrared, he'd have to select maybe one hundred of them and ask for time on an infrared telescope—assuming he knew how to use it—and spend weeks or months observing and analyzing. Now, in his office, he takes a computer file with 10 million stars from the Sloan and another computer file with 1 million stars from 2MASS (2MASS is less sensitive than the Sloan, so fewer stars show up in the infrared sky than in the optical). Then he uses software he's written to tell his computer to match the stars by position, so when he effectively finds an optical star on top of an infrared star, he knows they're the same star. And then he asks his computer to find the ratio of optical to infrared light for the majority of the matched stars, and then he knows what it is to be normal. Then he says to his computer, "Give me all abnormal guys. Then if you're weird, if you're one in ten million, you must be something interesting."

Nobody knows just how much of this jumping around in wavelength archives people are doing, but it's a lot. The publications list of the Sloan shows handfuls of papers comparing objects in Sloan with the same objects in other archives. A lot of the papers are about quasars, which in optical wavelengths look like stars, but which are bright in all wavelengths. The NSF's Wayne van Citters attends meetings put on by the NSF's own postdoc program and says that increasingly, a virtual observatory is just the way young astronomers think. "It's perfectly natural to them," he said. "They write software, they go to archives, they'll latch on to whatever they can find to put all the pieces together. And if it means that they don't go to the telescope, they think, Well, OK, so I don't have to go to the telescope."

Astronomers don't have to be hired only at universities rich enough to buy their own telescopes, they don't have to compete for time at oversubscribed telescopes, they don't have to worry they're intruding on claimed territory, they don't have to find data someone is hoarding. They can sit at their computers wherever in the world they happen to work and ask the archives whatever questions happen to occur to them. What's happened to astronomy, Van Citters thinks, is really, really interesting.

Another group using the archive is nonastronomers everywhere with access to the internet. Alex's postdoc, Andy Connolly, worked with Alex on Sky Server archive and then got a faculty job at the University of Pittsburgh. During a sabbatical from his job, in 2006, he went to Google to lead a project called Google Sky. Google Sky is a pannable, zoomable map of the sky that starts with the Sloan images, and where the Sloan doesn't cover the sky, uses images from a digitized Palomar Observatory Sky Survey. Run the cursor over the sky and the RA's and dec's show up. Select images in a list of wavelengths—the optical come from the Hubble Space Telescope, in X-rays from NASA's Chandra satellite, in ultraviolet from GALEX, and in infrared from NASA's Spitzer infrared space telescope—and see the image surrounded by its neighboring sky, in context. A little box pops up and tells you all about it.

WikiSky does much the same thing—run your cursor over the sky and the names and facts pop up—again, using the Sloan and much the same surveys. On WikiSky you can also see what the whole sky looks like in different wavelengths. The infrared sky is dominated by a scary-looking ring of heat and dust that follows the plane of the Milky Way; the ultraviolet sky has the same ring but it's black and the only light is a thin haze of stars forming outside the plane. Your first impression, panning around the Wiki or Google universe, is how much of it is black and dusted with undistinguished white dots. Microsoft's WorldWide Telescope is, as you'd expect, much fancier: it has everything the others do, plus narrated stories with data under the images. If you want to tour the universe with musical accompaniment, you can download recordings from Alex's Panta Rhei.

Galaxy Zoo is altogether different—no zooms, pans, pop-ups, or narratives. In 2007, at Oxford University, a postdoc named Chris Lintott and a graduate student named Kevin Schawinski each needed for their own reasons some blue elliptical galaxies, rarities because blue galaxies are young star formers and ellipticals were known to be all old, red, and dead. Schawinski had a small sample of blue ellipticals that he had found by searching through fifty thousand galaxies in Sloan's archive by eye: computers, for all their multifarious talents, are no good at differentiating one shape from another, and the human brain is superb at it. But those fifty thousand galaxies had all been at one particular redshift, and both Schawinski and Lintott wanted to find the blue ellipticals at different redshifts.

Lintott knew of a NASA project called Stardust@Home, which had captured dust grains from the tail of a comet, brought the dust back to earth, and embedded the grains in a gel. To find out whether any of those grains looked unusual and therefore might have come from outside the solar system, in August 2006 they posted images of the dust grains on the Internet and asked people to look through them. By September 2007, 24,000 people calling themselves "dusties" had looked through 40 million pieces of dust. Lintott thought, "If people will look at dust, surely they'll look at our beautiful galaxies."

He got together a small group of other astronomers who also needed galaxies classified by shape, one of whom was Bob Nichol, who called Alex Szalay and got him to donate space on a Sloan computer at Hopkins. The group put together a website whose first page is a brief tutorial on the shapes of galaxies and the point of the research. Then you're shown a galaxy and you click on a classification, spiral or elliptical; next galaxy, next classification. They called the website Galaxy Zoo. So far, it hadn't cost a penny; in about three years, they hoped, they would have ten classifications per galaxy, enough to be statistically convincing—science often has long lookout times.

To find the people to do the clicking, they put out a press release. And Wednesday morning, July 11, 2007, Lintott was interviewed on BBC's *Today*, his three minutes sandwiched between a segment on female circumcision in Africa that had run overtime and an interview with the prime minister, Gordon Brown. Afterward Lintott went off to a conference at the Royal Astronomical Society. Meanwhile, the BBC put the story on their website and the Associated Press had picked it up, as had several blogs including Slashdot, a website that calls itself "News for Nerds." Around noon, Lintott opened his laptop to see how the Galaxy Zoo website was doing, but he couldn't get through to it. So he opened the Galaxy Zoo e-mail and found more than ten thousand e-mails, mostly from people complaining that they couldn't get to the website. Back at Hopkins, in the predawn, the Sloanie in charge of the computers was woken by an alarm that went off when computers died. He went to campus to investigate and found that the computer Galaxy Zoo was using had overheated, its wires melted and fuses blown. He assigned another computer to the website, and by the end of the day, 22,000 people had done 500,000 classifications. By the next day, they were classifying in one hour the 50,000 galaxies that Kevin Schawinski had taken a week to do—the unit of classification became the Kevin-week. By the end of the week, Galaxy Zoo had done what its founders thought would take three years. By the end of the first year, Galaxy Zoo had 150,000 people, "zooites," who had done 50 million classifications, each galaxy classified for certainty more than thirty times.

The zooites are: a family physician, a truck driver, an owner of a Dutch computer business, a single mother, a Belgian secretary, a British sixth former, an art teacher, a Canadian college student, a supervisor on an oil rig, an actress, an architect, an astronomer's wife, a librarian, a medical student, a high school student, a mobile home park manager. A retired helicopter pilot gets up in the morning, feeds the sheep, gets a cup of coffee, checks the news, and classifies a couple hundred galaxies. A zooite got home from work one night, fired up Galaxy Zoo, and let his five-year-old watch him classify galaxies. The five-year-old asked if he could play the galaxy game again tomorrow. The zooite said of course but it wasn't a game, it was science, and the five-year-old said, stunned, "We're doing SCIENCE?" The zooites are primarily English speaking, but the only parts of the world that don't have at least a few of them seem to be deserts and jungles.

Lintott, Schawinski, and the rest of the group thought they should answer the zooites' considerable number of e-mails but obviously couldn't, so they set up the Forum, to which zooites sent questions, comments, weird things, general impressions. The Forum's moderators, the zookeepers—first Lintott, Schawinski, and the rest of the UK group, and then the increasingly experienced zooites themselves—answered questions about everything conceivable, and the zooites began getting an astronomical education. One zooite, a Dutch primary school teacher named Hanny, found a blue blob the zookeepers couldn't identify, and after a year, an international team of astronomers, including Hanny herself, got time on a telescope and found it was a jet of particles flying out of a massive black hole and slamming into a cloud of cold gas. Another zooite found a small, round, green object and claimed it was the first cosmic pea, until other zooites found others. Some enterprising zooites got into SkyServer and found the peas shared a certain spectrum—the zooites are seriously educating themselves about spectra—and are now looking for them at different redshifts. They've found 251 altogether. A Yale astronomer investigated further and found that the Green Peas—now officially capitalized—are small galaxies, a fraction of the Milky

Way's size, but are forming stars faster than almost anything else in the local universe.

One of Schawinski's undergraduates wanted to study the galaxies that could be classified as neither ellipticals nor spirals, a small grab-bag category called irregulars, so Schawinski asked the zooites to go find the irregulars. Then the undergraduate, as undergraduates do, gave up, and since none of the zookeepers were particularly interested in irregulars, the zooites took over. Before Galaxy Zoo, astronomers knew of 161 irregulars; since then, the zooites had collected 15,000 more. Lintott and Schawinski taught the zooites some astronomical analysis techniques. In order to use SkyServer, the zooites taught themselves SQL, which they said stood for "Slippery Quail Language." Irregulars, they found, are at all redshifts, but most are nearby; regardless of redshift, the smallest galaxies are likely to be irregular; and they're young, blue galaxies. The irregulars are probably Rich's faint blues. The irregulars' project, called Do It Ourselves, is working on a scientific paper.

So far, Galaxy Zoo has resulted in sixteen scientific papers, including one with the lead authors Schawinski (who finished his PhD and went to a postdoc at Yale) and Lintott saying that their 204 certified nearby blue ellipticals are about 5 percent of all nearby ellipticals, tend to live in more rural neighborhoods than normal red ellipticals, and that they are indeed forming stars, some of them furiously, that they show evidence of having active black holes, and that nobody can quite account for them. Maybe two blue spirals merged recently and formed an elliptical, or maybe they're red ellipticals in the path of some intergalactic gas that was shocked into forming stars, or maybe they're something else.

At last count, the zooites numbered 272,265. An astronomer named Thomas Wright, born in Durham, England, in 1711, was also, in the Enlightenment tradition, a mathematician, a builder of instruments, an architect, a tutor, and a designer of gardens. He was the first to figure out that we see the Milky Way as a band across the sky because we are seeing it from the inside. When he retired, he returned to his home village and built a small astronomical observatory. "I own I can never look upon the stars," he wrote, "without wondering why the whole world does not become astronomers."

Chapter 11

Everything's Different

It's beautiful. It's a stunning success and it was a very
innovative idea. It's changed the way we do astronomy.

—Matt Mountain, director, Space Telescope Science Institute

B Y 2001, the year after the Sloan officially began, the
survey was already in a standard textbook. By the end
of 2002, Sloanies had written 215 scientific papers with
Sloan data. By the end of 2003, *Science* magazine's Breakthrough
of the Year was the new standard model of the universe revealed
by comparing the cosmic microwave background as measured by
NASA's WMAP satellite with the large-scale structure as mapped
by the Sloan. By mid-2004, Sloanies had written 400 papers, and
non-Sloanies using Sloan data another 125. In August, Scot Klein-
man, an Apache Point observer, went to the Fourteenth European
White Dwarf Workshop in Germany and reported that nearly 40
percent of the talks mentioned the Sloan. A non-Sloanie attending
an American Astronomical Society meeting said he was astounded
at the way the Sloan permeated all the talks. In 2005, the American

Museum of Natural History in New York set up what amounts to an online documentary about the Sloan. In 2001 and 2006, of all the optical observatories—the Hubble Space Telescope included—the Sloan was the most productive; in the intervening years, no one bothered to rank observatories. "Not bad from a 2.5 m telescope!!!" Jerry wrote. As of October 2009, 2,656 papers were based on Sloan data and were cited in other papers 100,000 times. A non-Sloanie at the Space Telescope Science Institute said that Sloan hadn't even been on his radar, and now it was astronomy's eight-hundred-pound gorilla.

The Sloan had been promised as a survey of 10,000 degrees of sky in five years. Year One began in April 2000—though the Sloanies were too busy with hardware and software disasters to announce an official commencement—so Year Five was over in June 2005. In Sloan's very last scan, it found a star of the kind that Jill Knapp studied in her pre-Sloan days, an extremely massive star called an asymptotic giant branch star, normally too bright for the CCDs, but this one was obscured by a shell of dust. Jill felt she'd been waiting seven years for this star, but by the time of that last scan someone else on another telescope had beat her to it.

By mid-2003, halfway through the official run of the survey, they knew they weren't going to finish 10,000 degrees of sky. The science factory was finally working smoothly, but the Sloanies had underestimated the Apache Point weather, which had gone through a long spell of being unusually bad, and therefore overestimated the rate at which the data would flow. The part that wasn't getting done was actually the easiest, the sky straight overhead, in the Northern Galactic Cap. Michael Blanton, a Sloanie at New York University, called it the observing sweet spot; Jim said the need to finish it was an excellent funding tool.

So the Sloanies proposed what became Sloan II: a project called Legacy to complete the Northern Galactic Cap; one called SEGUE, for Sloan Extension for Galactic Understanding and Exploration, to figure out what was going on with the stars streaming in the Milky Way's halo; and a third, the SDSS Supernova Survey, to learn more about the universe's acceleration, dark energy in particular. Sloan II began officially in July 2005, and the collaboration grew to twenty-five institutions.

Almost immediately they began thinking about Sloan III, four projects this time, subsurveys, each one aimed at specific science. BOSS, the Baryon Oscillation Spectroscopic Survey, would continue to collect LRGs to refine the standard ruler that would help constrain dark energy. MARVELS, the Multi-object APO Radial Velocity Exoplanet Large-area Survey, would set a record for jerry-built acronyms and hunt for planets around other stars, partly for the sake of the planets themselves, partly to figure out how solar systems form. SEGUE-2 would continue the history of the Milky Way. APOGEE, the Apache Point Observatory Galactic Evolution Experiment, would build a spectrograph that operated in the infrared to see through all the clouds and dust in the plane of the Milky Way that Sloan I had to avoid. With SEGUE and APOGEE together, astronomers would begin to understand the Milky Way as a messy, nearly biological complex of stellar motion and chemistry; that is, they could understand at least one galaxy from the stars up.

No assumptions were made about Sloan III's membership, and any old or new institution wanting to join would pay $900,000; about twenty did. Princeton rejoined, of course. So did Hopkins. Chicago didn't rejoin, though Don York and Rich Kron stayed on: how could you walk away from the project you conceived and delivered? Fermilab didn't rejoin either. Its ex-Sloanies, John Peoples in particular, had turned their attentions to a new Dark

Energy Survey, and young Fermilab physicists who already knew how to write software to find luminosities and distances apart were now reading astronomy books and promising they'd "study the crap out of the sky."

Sloan III is run by the younger generation. Rich Kron handed directorship over to Daniel Eisenstein, the former Institute and Chicago postdoc. Some Sloanies now called themselves Threeons, a reminder that one path into astronomy has always been through science fiction. At Sloan III's first general collaboration meeting in July 2009, roughly half the Threeons were new, either because their institutions had just joined or because they themselves were new postdocs or graduate students. Daniel reminded the Threeons new and old that when they're in the shower and think of some new science to do, right after they've toweled themselves off, they should announce the project on the e-mail list. Daniel was reminding them of the rules of their new culture.

The Sloan, in the roughly twenty years from its beginning until Sloan III, changed the way astronomy is done. Everyone says so, and in almost exactly those words. Before Sloan, astronomers were like writers: you think of an idea for a book, you research it, you decide what goes in and what's left out, you decide the story; it's yours, you own it. Science and books both have always existed independent of their authors, but they still carry in themselves the shapes of their authors' hands and minds, they embody their authors. But when someone—Jim, Rich—wants to write not a book but an encyclopedia, everything changes. The project is too big for one person or a small group of people, so a collaboration must form, a community. And in that process, the emphasis shifts from the individual to the community.

But shifting to a community for an astronomer is tricky. Their culture and its rewards—publications, jobs, grants, prizes—were based on individual brilliance. All the agony that the collaborative

Sloan postdocs went through still obtains—no science means no jobs and no NSF grants, and when you finally do publish, your publications have 150 co-authors. The Sloanies probably worked this out as well as it can be worked out. Their Principles of Operation, designed to spread the wealth and give the credit, said that anyone could work on any science and be on any publication to which they contributed. The NSF still doesn't quite know how to deal with this shift to community that surveys entail but is aware of the issue and is thinking it through.

And the problem isn't going away, because after Sloan, large digital surveys seemed feasible and desirable. By the time of Sloan III, several other surveys were in the works, all of them offering superlatives—largest, widest, deepest—and huge numbers. Sloan used a 2.5-meter telescope, a camera with 120 megapixels, and produced an archive with 40 terabytes of data. Pan-STARRS, Panoramic Survey Telescope and Rapid Response System, is a telescope with effectively four 1.8-meter mirrors and four cameras, 1.4 gigapixels apiece, that will produce several terabytes a night. LSST, the Large Synoptic Survey Telescope, is an 8.4-meter telescope with a 3,200 megapixel camera, 30 terabytes of data per night, and when the survey is over, the archive will hold 60 petabytes—"peta-" means "quadrillion." Pan-STARRS has one main parent, the University of Hawaii's Institute for Astronomy, and several collaborators, including Johns Hopkins; each institution owns a particular part of the science, and once the survey is finished the archive will be made public. LSST's collaboration is enormous and includes Caltech, Google, Hopkins, Princeton, the University of Washington, and about thirty others. The data will be public.

Jim, trained by great professors and practicing for years as one, likes what he calls this "team organism." He thinks working this way is fun: "You're not only proud of yourself," he said, "you can be proud of the people that are working with you, they can be proud of you." A further niceness: shifting from one astronomer to a community then shifts the emphasis even more, to the thing being made, the encyclopedia, the survey. It's a thing separate, external to its authors and outweighing anything their hands could do or their minds could conceive. Its creators feel as though they're part of some great thing, as Rich Kron had said, some grand and bold thing.

Besides, so far, it worked out well for the young people. The Sloan graduate students got good postdoc positions, the Sloan postdocs got good jobs. Some of them suspected they all got the jobs they would have gotten even if they hadn't spent so much time on hardware and software; others weren't so sure. But as more and more institutions joined the collaboration, more and more astronomy departments fielding Sloanie applications knew the postdocs' names and understood what they'd done.

David Hogg, who had been an Institute postdoc, and Michael Blanton, who been a grad student at Princeton and a postdoc at Fermilab, both got faculty jobs at New York University. Daniel Eisenstein, an Institute and Chicago postdoc, went to the University of Arizona; later, so did Michael Strauss's student Xiaohui Fan. Doug Finkbeiner, a postdoc at Princeton, got a faculty job at Harvard. Heidi Newberg, a Fermilab postdoc, got one at Rensselaer Polytechnic Institute, and Michael Richmond, a Princeton postdoc, at Rochester Institute of Technology. David Schlegel went to a senior position at the Lawrence Berkeley National Laboratory, where he worked with Nikhil Padmanabhan, a postdoc who had been a graduate student at Princeton; later Nikhil joined the faculty at Yale. Jim Annis stayed on at Fermilab as a scientist. Connie Rockosi, who had joined the Sloan as an undergraduate at Princeton, went to graduate school at Chicago and did a postdoc, for which she received a prestigious Hubble fellowship, at

UW, and then in her mid-thirties, having spent half of her life on the Sloan, got tenure at the University of California, Santa Cruz, whose astronomy department is unusually sympathetic to instrumentalists.

On January 9, 2008, a freshman at Princeton named Jared Crooks presented a poster at a meeting of the American Astronomical Society, in a "ginormous room," he said, "so it was kind of overwhelming, and I wasn't used to that." Crooks is black; black astronomers are more rare than brown dwarf stars. He had gone to the High School of Medical Professions in Fort Worth, Texas, a public magnet school that had a hard time meeting federal testing standards and offered neither advanced physics nor calculus. The school board bought him two textbooks so he could learn on his own. Crooks thought he wanted to be a cardiologist until he watched open heart surgery and got bored. He was easily bored; he felt he could want something, maybe like a Ferrari, but if suddenly he could have it, he'd love it for a week and after a month he'd know the Ferrari and be bored. Then one night, he came home and got out of the car, an especially clear night, and he looked up. Two hours later and he found himself still looking. "They look beautiful," he thought, "the stars are beautiful." Whatever that beauty was, he wanted to do something with it and he'd never know all of it; the sky was a changing Ferrari.

He went to summer programs for three years at an East Coast private school training minorities to go into math and science. He thinks that's how he got into Princeton. The summer before school started, he wrote to Jill, who directed the department's undergraduate studies, and asked for a job. Jill hired him to write computer code to compare Sloan data with data from a radio survey. He was excited at first but once he started, he felt lost and a little frustrated that he wasn't getting at the stars by writing code. But then the program worked and he started getting results, and when he found a supernova no one had known about and then when he presented his poster at the AAS meeting, he thought, "OK, this is what I'm doing now. I was just tying my shoes before I started running. Now, OK. This is what I'm doing."

The best reward for working in a community is what you can do with terabytes. In the late 1990s, Michael Blanton was a graduate student at Princeton, working with Michael Strauss and Jerry Ostriker on a theory of galaxy evolution, in particular, on confronting the theory with observations. Blanton would read a published paper based on 150 or even 1,000 galaxies, selected in some particular way, using particular definitions, and observed with some particular instrument that had its own vagaries. The paper would be exciting, and he'd say to himself, "Wow. Let me compare it to this other paper." And the other paper would be based on a few hundred different galaxies measured with different selection criteria on different instruments, and its graphs would plot different characteristics. So he'd try to figure out whether the papers were agreeing or disagreeing, and he couldn't always tell.

Blanton finished his PhD and in 2001 began a postdoc at New York University with David Hogg, who was a professor there. Hogg had liked the freedom he'd had as an Institute postdoc, so he advised Blanton to work on whatever he wanted to; luckily, Blanton wanted to work on Sloan galaxies with Hogg. Before Sloan, most galaxies were known to come in two basic shapes, spirals and ellipticals—the tutorial given to the Galaxy zooites. Spirals were known to live in the universe's rural fields and to be blue because they were forming stars, which are born in a hot shower of ultraviolet light. Ellipticals were known to live in clusters and to be red because they were full of dying or dead stars whose light had

cooled to red or infrared. A reasonable theory was that the cluster ellipticals had been born cosmologically long ago as spirals,
then merged to become ellipticals, which aged into redness, and
the field spirals were born cosmologically recently. But with small
samples of tens or a few hundreds of galaxies, the theory wasn't
convincing.

Then in 2001, Sloanies took around eight hundred galaxies, classified their shapes by eye and showed that sure enough,
they came in two types, blue spirals and red ellipticals. In early
2003, Tim Heckman, his graduate student Christy Tremonti, and
Sloanies that included a group from the Max Planck Institute for
Astrophysics in Garching took 100,000 galaxies and plotted the
added-up mass from all a galaxy's stars against the rate at which the
galaxy was forming stars and found that low-mass galaxies were
young and forming stars, and high-mass galaxies were old and not.

That same year, Blanton and Hogg did a simple and obvious
thing. They took the images and spectra of 100,000 galaxies and
corrected for their redshifts so that all 100,000 galaxies seemed to
be at the same distance. Then they plotted every galaxy characteristic against every other characteristic: color (blue to red) against
brightness (bright to faint), size (big to little) against the concen-
tration of light in the galaxy's center, which—since ellipticals have
denser brilliant centers than spirals—is a proxy for shape. Arrange
the characteristics in a matrix so that color is also plotted against
size, size against brightness, brightness against shape, and so on.
No matter which characteristic was plotted against which other
characteristic, the graphs showed two clumps, two populations of
galaxies. Galaxies were either big and red and bright, or they were
small and blue and faint—the former tended to be ellipticals and
the latter, spirals—and not much was in between. Cosmologists
call this bimodality.

Next, several other groups plotted colors of the galaxies against the crowdedness of their neighborhoods and found the red ones were in crowded neighborhoods and the blue ones liked the suburbs or the fields. And like the characteristics Hogg and Blanton plotted, all these characteristics too were bimodal; one or the other, not much in between. Red galaxies are large, bright, massive, elliptical, urban, and not forming stars. Blue galaxies are smallish, fainter, less massive, spiral, lonely, and forming stars like fireworks.

In a huge data set, Jim says, you can look for relations between characteristics, and once you have the relations, you have something to explain. What needs explanation in bimodality is why those particular sets of characteristics go together. Why are red, non-star-forming galaxies also big and bright? They're bright because, as Jim says, they're a *hell* of a lot bigger than the faint little blue star formers. Then why so big? Cosmologists had liked the merger hypothesis: little things form first and merge into bigger things. So then why weren't the little things red and dead and the big things blue and lively? The reason had to be that the big things formed first. Or rather, maybe they were little when they formed but because they were in the neighborhood of other galaxies, they merged and got bigger. In any case, the point is that when a galaxy starts early, it gets big. The galaxies that are little now were born late. Cosmologists are calling this "downsizing," and Jim says the idea has only slowly crept into their brains.

The next question is whether, as seems reasonable, the two populations are different stages in a single life history, adolescents and seniors of the same species, and the answer to that is less clear. Galaxies begin in the blue population, and as their stars form, age, and die, they move in general to the red one. But they don't seem to move gradually, because none of the bimodal graphs have much between what they called the blue cloud and the red sequence, so galaxies don't move from blue to red gradually; they never look middle-aged. That means, says Jim, that something happens that stops you forming stars, and whatever it is, it's abrupt and dramatic and affects the whole galaxy. Figuring out the physics of that process, says Jim, is not impossible but is very, very hard and hasn't been done yet. In any case, the real difference between the populations is less that they're spirals and ellipticals than that they're star formers or they're not, alive or dead.

The next thing to do will be to make those same graphs— everything against everything—at a series of different redshifts and watch the populations changing with time. So far, teams of non-Sloanies with access to huge telescopes and deeper surveys but smaller numbers of objects have found that going back in time, galaxies are indeed bluer; coming forward in time, they're redder; and the trip is only one-way, blue to red. So the red population, like the populations in cemeteries, will do nothing but grow.

As with the baryon acoustic oscillations, the Sloanies couldn't have found bimodality if the survey had been done more sensibly. The Sloan spectroscopy could have been done with exposures of minutes, long enough to measure galaxy redshifts well enough to map large-scale structure. But from the beginning, Jim had wanted "real spectra," he said, and so the exposures were closer to an hour. And the survey, instead of taking one year and, said Hogg, producing twenty papers, it had taken seven years and produced two thousand.

The bimodality of galaxies had been talked about for years in a hand-waving manner by astronomers with small samples. Jim thought Hogg and Blanton used the full power of the survey and, he said, "just nailed the problem." Finding bimodal populations isn't the kind of discovery, like the discovery of cosmic acceleration, that explodes the standard science. It's the kind that infiltrates the science, tightens its focus, convinces the scientists that these incredible theories about an incredible universe are credible after all. Bimodality was the kind of problem Jim had in mind for the survey from the beginning: "ecology on very large scales," he said, systematic and precise observations of relationships that made sense of the universe as a whole. In the Sloan archive, the universe came closer to being one coherent story.

If the Sloan culture changed the way astronomers work, then Sloan science has seemed to change the way they see the universe. Sloan science covered an enormous number of disparate fields. Maybe because the data from all those fields were all in the same archive, maybe because the Sloan name and the same Sloan authors appeared on so many papers on such different subjects, maybe because the conference at the end of Sloan I covered every field on every scale—whatever the reason, the Sloan gives you the impression that for the first time, the universe looks like one interlocking, interrelated, cause-and-effect system.

Chapter 12

Jim Again

But if the matter were evenly disposed throughout an infinite space, it could never convene into one mass; but some of it would convene into one mass and some into another, so as to make an infinite number of great masses, scattered throughout all that infinite space.

—Isaac Newton

It's easier if you're smart, but I don't think it's absolutely necessary. I think dedication really matters more.

—Jim Gunn, Princeton University

S TART NOW NOT with scale, small to large, but with time, at the beginning—though not at the very beginning, which is the province not of cosmology but of theoretical physics, some of it highly theoretical. Cosmologists' beginning is around 100,000 years after the Big Bang, around a redshift of 3,000. The universe had three actors, light and matter and gravity. Light is all wavelengths of radiation and can be thought of either as a wave or as a particle called a photon. Matter is of two kinds: the protons and electrons that make up ordinary atoms and the extraordinary particles—no one knows what they are—that outweigh the ordinary ones by six to one and that are called dark matter. And gravity is gravity: it pulls on everything, dark and ordinary matter alike, and the bigger something is and the closer you are to it, the more strongly you're pulled.

At a redshift of 3,000, matter and light were in a kind of uproarious equilibrium. The universe was expanding 120,000 times faster than it is now, and was 8,000 K; it was a plasma so furiously hot that the matter, the protons and electrons and dark matter, all raced around freely. The photons of light traveled with them; they couldn't travel alone or very far in this dense universe before electrons would scatter them off in other directions, and scattered photons meant the universe was opaque.

As the universe expanded and cooled, the plasma fluctuated; it had waves in density and temperature, little pile-ups like hills, some hotter, some cooler. Gravity would have liked to make the cooler, denser hills bigger, but with ordinary matter, it couldn't: at these temperatures, the photons in the hills carried enormously more energy than the protons and blew the protons out of the hills into rings, shells around the hills, little ripples that spread out farther and farther, so small—only one part in 100,000—that they hardly disturbed the universal equilibrium. Those energetic photons, however, couldn't touch dark matter, so gravity was free to pile the hills of dark matter higher, then higher yet, up into mountains. And the universe kept expanding and cooling, dark and chaotic.

When the universe was around 400,000 years old, at redshift 1,000, it dramatically shifted state. The temperature had fallen to 3,000K, and photons had now lost enough energy that they could no longer interfere with protons. Electrons and protons could attract each other and bind themselves into the first elements, the atoms of hydrogen; cosmologists call this time recombination.

With no electrons unbound, the photons no longer scattered but streamed freely, and suddenly the opaque universe became transparent. The photons of light traveled on alone, cooling, dimming, their wavelengths lengthening with the expanding universe, and played no further part in its history until 1965, when scientists found almost by accident a background of light still filling the universe but lengthened to microwaves and cooled to 3 K, 3 degrees above absolute zero.

And now, after recombination, the universe was a gas of hydrogen atoms. The mountains of dark matter were still growing, connecting up into chains. All around the dark matter mountains, the ripples of ordinary hydrogen are still spreading, overlapping with other ripples. All together the ordinary and dark matter have laid down a shadowy pattern of lower and higher densities. Along this pattern, gravity begins to win, acting on dark matter and ordinary hydrogen alike. Gravity pulls the hydrogen into the mountains of dark matter, pulls the dark matter into the rings of hydrogen, mixes the dark and ordinary matter together. Gravity pulls those mountain chains and rings in on themselves; they contract, fall together until they stop expanding with the universe, and begin collapsing. Small things collapse sooner, bigger things later, a whole hierarchy of structures in slow collapse.

The universe still expands, dimming, glowing more and more redly as hydrogen absorbs what light is left, and darkening. Along the pattern laid out in dark matter and hydrogen, small clouds condense, merge together, swirl into shapes that might someday look like spirals. The dark matter can't cool off enough to keep condensing, and it collapses just so far, by about half. The hydrogen gas collides with itself and heats up, radiates the heat away, cools and condenses. It could collapse forever, but it's got a little spin, then a larger spin, and centrifugal forces hold it against further collapse, and it forms a spiral disk.

Somewhere around redshift 10, here and there in the emerging spirals, the hydrogen has cooled enough that it becomes dense enough to make stars. In the stars' cores, the hydrogen atoms fuse together, heat up to 10,000,000 K, create heavier elements, and in this process of thermonuclear fusion, shine crazily, throwing out floods of ultraviolet light. In the same neighborhood and at the same time, some of these spirally clouds full of stars and gas collide and merge and become much bigger, much denser, and more nearly spheres. In the spheres, shocked gas falls into brilliant, ultraviolet stars. The neighborhood is full of these messy, gas-rich things, orbiting one another, falling together, lighting up into stars. The darkened universe is dotted with points of brilliant ultraviolet.

The ultraviolet radiation pouring out of the stars hits the hydrogen in these protogalaxies and in the neighborhood around them. Every time an ultraviolet photon hits an atom of hydrogen, it blows the atom's electron off, leaving the atom ionized. As more and more stars form, more and more of the neighborhood's hydrogen becomes ionized, and since the same thing is going on in all other neighborhoods, gradually the ionization spreads. By around redshift 6, the whole universe is reionized. And because it's still expanding, the electrons have now gotten too far from their atoms to ever recombine with them and the universe is now ionized forever. The electrons are also small enough and far enough apart that the photons rarely encounter them and the universe is now transparent forever. New stars continue to light up, and the universe is bright with them.

Inside these spiraling and spherical protogalaxies, the hydrogen not fused into stars is pulled toward their centers, getting denser and denser until somehow—no one knows how—gravity wins completely. The hydrogen's mass collapses to a single dense point, a singularity, at which gravity is so strong that light itself is caught by the singularity, so that at the centers of the protogalaxies are black holes. The protogalaxies are still merging and swapping gas which also falls to the centers and just before entering the black holes is ripped to atoms. It shines excitedly and makes a quasar, one quasar per galaxy, and they too pour out ultraviolet, and the universe gets even brighter.

The protogalaxies now shape up into galaxies. The spirals have large, bright centers and wheeling arms traced out by the ultraviolet and blue of forming stars. Neighboring clouds of gas and small protogalaxies still fall into them, though nothing too large or the delicate spirals disrupt. The ellipticals can handle any amount of infall and only get denser and brighter and bigger; the earlier they're formed, the longer they have to grow, the bigger they get. None of the galaxies grows much larger than a trillion times the sun's mass, and no one is sure why. The neighborhoods gradually condense into clusters of galaxies, lining up along the shadowy pattern—still one part hydrogen to six parts dark matter—laid down at recombination. Because spirals are so easily disrupted, they more often disappear and get recycled into ellipticals, so the clusters gradually fill with ellipticals lighting up with stars. By now, the redshift is 2.5 and galaxy formation is at its peak; it's the epoch of quasars. The universe is nuclear blue, full of explosions, ignitions, collisions; it looks like a carnival in full swing.

In one of the less-interesting neighborhoods, the Milky Way has pulled together a center around which it's forming a disk where gas is turning into stars and falling into the central black hole, our own local quasar. A few nearby smallish, irregular galaxies are attracted to the Milky Way and fall into it, their stars and gas disrupted into streams that eventually spread out in a halo around the whole galaxy.

In the same neighborhood as the Milky Way and the irregulars is only one other sizable galaxy, M31. It's another spiral, but a little older and bigger, and like us, growing without fanfare. The neighborhood around both galaxies is pulling together into a loose cluster. And everywhere in the universe other clusters, lining up along the ancient pattern in the matter, are pulling together into superclusters, binding their galaxies to one another by mutual gravity. The rich get richer, and the thin places in between thin into voids.

Around 5 billion years ago, at a redshift of around 0.5, in one of the Milky Way's outer arms, our solar system formed. The sun condenses out of cooling gas that, like the Milky Way, spins into a disk in which planets accumulate themselves, and whatever is left over breaks apart and circles the whole system as families of asteroids.

Somewhere between redshifts 0.5 and 0.2, the universe has become tenuous enough, the density of matter low enough—nobody knows why at all—that its infinitely long, gradual deceleration changes sign, gravity loses the fight against some kind of repulsive force. Dark energy kicks in, and the universe accelerates.

Throughout, gas keeps turning into stars and fueling black holes, but more and more slowly now because the universe has only so much gas and it's all getting farther apart faster. Eventually the gas is rarefied enough that the quasars' black holes gradually go silent, sitting in their galaxies brooding. And stars, once born, age. Depending on their masses, they have lifetimes that range from tens of billions down to a million years and in any case, are limited. The little red stars live a long time. The big blue stars die young. So slowly, as blue stars burn out, galaxies that began blue turn red. And between and within the galaxies, only a little gas is still around to pull in and kick up more stars. Once a galaxy is red, it waits only to burn out entirely. The large red ellipticals, the LRGs, are at least useful for tracing out the ancient pattern of rings, now 450 million light-years across, around the mountain ranges of dark matter, now 300 million light-years long, and that together outline the large-scale structure. As the gas turns into stars and the stars burn out and the gas runs out, the universe keeps flying apart, faster and faster, and everything keeps getting colder and darker and farther apart. The universe won't end at all, but nobody will care.

This science is not by any means the final truth. Nor was it all done by the Sloan, but much of it was. Jim did almost none of it. He advised, he suggested, he discussed for hours, his name is on the papers, but he's never extracted a piece of data from the archive and has no idea how to do it. He could learn and quickly, but he hasn't. And not having done Sloan science, he's not likely to start: the median age of astronomers is in their forties, and they start leaving the field in serious numbers in their mid-sixties. In 2008, he turned seventy. Would he like to do some astronomy? "Yes, of course I would," he says. "But it's been a long time since I've done any, and I'm not as young as I used to be. There is this nagging question —I was very good once—and the question is, would I still be so good if I did it again?"

In fact, Blanton and Hogg asked him to collaborate with them on their work on elliptical galaxies and their neighborhoods, work that they felt followed up on Jim's earlier work. And Jim, though interested and on call as a consultant, never got around to it, he said; just too busy. Besides, Hogg and Blanton had written their own software to analyze the data, and Jim didn't have time to learn their software. "The tools to do this stuff sort of passed me by," he said. "I wasn't in any position to contribute in a material way to this research because I just didn't know how." Jim built the survey and that counts, he thinks, but he hasn't done any real work, and he feels that's a sadness, in fact, an immense sadness.

Some Sloanies, in their spare moments, have thought the same thing. Tim Heckman, working with 100,000 galaxies on the relation between their masses and star-forming rates, thought that this was the sort of thing Jim would have wanted to do and felt a little sad that Jim didn't get to have the fun. David Hogg, who thought that Jim had written four or five of the greatest papers in astronomy but then scientifically disappeared, quoted David Spergel, a non-Sloan astronomer at Princeton, saying that the Sloan was great but ten more years of Jim Gunn would have been too.

Jim is no longer Sloan's project scientist; with Sloan III, he turned the job over to David Weinberg, a long-time Sloanie who'd been Jim's graduate student, an Institute postdoc, and was now on the faculty at Ohio State University. Sloan III's subsurveys have a lot of hardware in common, much of which has to be built. They call it infrastructure, and Jim is now the infrastructure lead, in charge of hardware/operations interfaces. Jim's camera, needed only for the first two years of Sloan III's BOSS project, was taken off the telescope and off the mountain, and Jim isn't sure what to do with it. He thinks it would make a good coffee table.

Toward the end of Sloan II, the Sloanies gave themselves a congratulatory symposium called Asteroids to Cosmology, four days of talks by Sloanies and Sloan data users on every kind of science in the fields of astronomy and cosmology: the solar system, stars, the Milky Way, galaxies, quasars, the large-scale structure, the small, the nearby, the astronomically large and distant, the whole universe. Jim was to give the conference summary, so he sat in the front row taking notes. Conference summaries are just that, some smart person with a wide education who knows everything listens to all the talks and sums them up so everybody knows what the conference has been about. It's always the last talk on the last day. So on the last day, Jim gave the summary, and when he finished, the audience began applauding and wouldn't stop. When they finally did, the session chair said it was the longest ovation ever at any conference. Then she said the pro forma, "To close the conference, let's thank our last speaker again." This time the audience applauded, then stood up and applauded some more and kept on applauding, a standing ovation that seemed as though it would never end. Astronomers always applaud, but they never stand. They just don't do that; they're a competitive and meritocratic democracy, and nobody's treated that much better than anyone else. The only other astronomical standing ovation anyone remembered had been for WMAP's results on the microwave background, but that was for science, never for a person. Jim stood there in his Hawaiian shirt and didn't say a word. He looked self-conscious, and his nose turned pink.

In 2005, Jim won three large prizes: the American Astronomical Society's highest prize, the Henry Norris Russell Lectureship; the Gruber Foundation's Cosmology Prize; and the Crafoord Prize given by the Royal Swedish Academy of Sciences—the same outfit that gives the Nobel Prizes in most sciences except math and astronomy. In 2009, he won the National Medal of Science. Every time Jim won a prize, Jill gave him a teddy bear, so added to the three prizes he'd previously won, Jim has a lot of teddy bears. He likes the Crafoord prize one best; it's a doughty Brit-looking bear, "A proper little fart," Jim says.

For the Russell prize, he had to give a lecture. He advised the audience to think carefully about the balance between theory and technology. Certainly cosmological theory has, in the last fifty years, become a triumph of human intellect, he told them, but we wouldn't know any more than we did fifty years ago without the explosion in technology, both hardware and software. The technology, he said, is every bit as hard and gratifying as astronomy, but "the astronomical community has not recognized this fact nor adequately rewarded the practitioners, and makes no effort to educate more." Everyone knows that new astronomy is always driven by the next technology. So if you don't want any more triumphs of human intellect, he said, continue to not reward hardware builders and software writers and continue to not train new ones.

Otherwise, Jim thinks cosmology is in pretty good shape. "We really know a lot," he says. "We know pretty much what the universe we can see looks like." Then being Jim, he worries about what we don't know, the part we can't see, what's called the dark sector, dark matter and dark energy. He's disconcerted that the universe is simple enough that you can sit in an armchair and put together statistics and the physics of gravity, radiation, and particles, and the rules you come up with match the operations of the visible universe. He's surprised that the Sloan, with its unprecedented precision and comprehensiveness, came up with nothing that didn't fit. Nothing Sloan has seen was unpredicted or unsuspected—nothing, Jim says, that would make anyone think the simple universe is wrong. He thinks that the completely unexplained nature of the dark sector ought to mean that we don't understand some fundamental things. And those unexplained fundamental things should have had some sort of effect or resonance or repercussions that Sloan should have found. Something should have not fit. That everything nevertheless did, seems to him to be—he uses all these words—incredible, fantastic, entirely remarkable. It disturbs him.

At the end of Jim's Russell talk, he said that the future was of course up to the young people in the audience. "Many of you are just beginning your first half century in astronomy. Most of the SDSS science was done by people who began work in the project as young as the youngest of you, and that has been very gratifying to me." He ended, "It is a wonderful, exciting field, which has been generously supported. Take care of it."

If you press Jim on the sadness of building the survey and not doing the science, he's pragmatic: "That's what I decided and that's what I've done," he says. "I've managed to work that around in my head in a way—it wasn't a conscious thing—that I can just feel fatherly about this. These are my children doing this wonderful thing," he says. "I'm happy to have had my part in making it happen."

Željko Ivezić grew up in Croatia, got interested in astronomy because his favorite girl in third grade was. He came to the United States because opportunities in astronomy were greater in the United States than in Croatia. He got a PhD from the University of Kentucky on the effects of dust on stars—"Like car lights through fog, and you want to know what kind of car it is," he said. He wrote now-famous software called Dusty to help astronomers correct starlight for the effects of dust. Because Jill also worked on stars and because she needed code writers, she offered Željko a postdoc at Princeton, which in 1997, he accepted. He worked with Robert on the photometric pipeline, staying there for seven years. Unlike Jim, he has done a great deal of science with the Sloan; he's been first author on eleven Sloan papers on subjects from stars to the Milky Way to asteroids and in the first tier of authors on about fifty more. In 2004, he got a faculty job at the University of Washington and not only continued to work on Sloan but also signed on to work on the next triumphal survey, the LSST. His title is system scientist, approximately what Jim's title had been on the Sloan. He spends a lot of time doing managerial coordination in teleconferences, which is not what he trained to do, nor does he enjoy it.

He does it nevertheless, he says, because "I observed it when I was young, people like Jim. They were excellent scientists but they sacrificed their science time to make Sloan happen. And so I think it's just now the turn of our generation, that we do a little bit of boring stuff so that generations that come after us, they will enjoy the new survey just as much we enjoyed the Sloan. Or as someone said—the purpose of life is to plant the trees whose shade we will not enjoy."

Acknowledgments

I WANT TO THANK: my editor, Emily Loose, for being kind and ruthless, and for loving the science; my agent, Flip Brophy, for being her tough and golden self; my colleague, Richard Panek, for listening patiently to whining and for reorganizing the prologue; and my husband, Cal Walker, for saying, "You always say that and it always works out." I especially appreciate the Sloanies for being so intelligent and open, for taking so much time, and for explaining, correcting, thinking, fact-checking over and over. The worst part is that most of them don't appear in the book.

Sources

THE MAJORITY OF the information in this book came from interviews (listed below) and from the personal papers and saved documents of the interviewees.

Much information also came from the archived emails of SDSS I and II, an archive that is private but to which I was generously given access.

The scientific papers described throughout the book can most easily be found on the SDSS website's publications list: http://www.sdss.org/publications/.

For a good interview of Jim Gunn, see American Institute of Physics, Center for History of Physics and California State University, Fullerton, Oral History Program. Modern Cosmology: An Oral History Inquiry. Perceptions of Scientific Works. James E. Gunn. Interviewed by Paul Wright, February 10, 1975.

Interviews

Done in person or on the phone since January, 2007, not counting innumerable e-mails for checking facts.

Jim Annis, Fermi National Accelerator Laboratory, Batavia, IL: 10/15/07.

Neta Bahcall, Princeton University; at Space Telescope Science Institute, Baltimore, MD: 6/21/07.

Tom Barnes, National Science Foundation; by phone: 5/24/07.

Mariangela Bernardi, University of Pennsylvania, Philadelphia, PA: 12/14/07.

Michael Blanton, New York University; at Sloan Symposium, Chicago, IL: 8/16/08.

John Bochanski, Massachusetts Institute of Technology; at University of Washington, Seattle: 4/11/07.

Bill Boroski, Fermi National Accelerator Laboratory; at Johns Hopkins University, Baltimore, MD: 12/6/07, 12/7/07.

Jon Brinkmann, Apache Point Observatory; by phone: 11/21/07.

Larry Carey, University of Washington; at Apache Point Observatory, Sunspot, NM: 7/6/07, 7/8/07, 7/10/07; at Johns Hopkins University, Baltimore, MD: 4/16/08.

Michael Carr, Princeton University; at Apache Point Observatory, Sunspot, NM: 7/10/07.

Hirsh Cohen, consultant, Sloan Foundation; at Sloan Foundation, New York, NY: 1/16/07; at Sloan Symposium, Chicago, IL: 8/17/08.

Jim Crocker, Lockheed-Martin; at Space Telescope Science Institute, Baltimore, MD: 11/13/07.

Jared Crooks, Princeton University, Princeton, NJ: 1/16/08.

George Djorgovski, Department of Astronomy, Caltech, Pasadena, CA: 3/7/07.

Tom Dombeck, University of Hawaii, Institute for Astronomy; by phone: 3/18/09.

Daniel Eisenstein, University of Arizona; at Johns Hopkins University, Baltimore, MD: 5/7/07; at Space Telescope Science Institute, Baltimore, MD: 5/7/08.

Michael Evans, University of Washington, Seattle, WA: 4/11/07.

Sandra Faber, University of California, Santa Cruz; at Sloan Symposium, Chicago, IL: 8/15/08.

Doug Finkbeiner, Harvard University, Cambridge, MA: 8/14/07.

Doug Finkbeiner, and David Schlegel, at Harvard-Smithsonian Center for Astrophysics, Cambridge, MA, 8/13/07; 8/14/07.

Wendy Freedman, Carnegie Observatories; at Sloan Symposium, Chicago, IL: 8/17/08.

Joshua Frieman, Fermilab and University of Chicago; at Drexel University, Philadelphia, PA: 3/30/07.

Russ Genet; by e-mail: 10/25/07–10/29/07.

Bruce Gillespie, site manager, Apache Point Observatory; at Baltimore, MD: 3/5/07; at Apache Point Observatory, Sunspot, NM: 7/6/07; by phone: 10/26/07.

Murph Goldberger, University of California, San Diego; by phone: 6/7/07.

Edward Guinan, Villanova University; at Space Telescope Science Institute, Baltimore, MD: 5/23/07.

Jim Gunn, Princeton University, Princeton, NJ: 5/14/07, 7/8/07–7/10/07, 8/17/07; at Johns Hopkins University, Baltimore, MD: 9/12/07; at Princeton: 1/15/08–1/17/08; at Johns Hopkins: 4/16/08; by phone: 5/20/08; at Princeton: 10/21/08–10/23/08, 7/16/09; by phone: 7/20/09.

Will Happer, Princeton University, Princeton, NJ: 8/15/07.

Tim Heckman, Johns Hopkins University, Baltimore, MD: 2/15/08.

Craig Hogan, University of Chicago and Fermilab; at Sloan Symposium, Chicago, IL: 8/18/08.

David Hogg, New York University; at Columbia University, New York, NY: 3/11/08.

Željko Ivezić, University of Washington; at Sloan Symposium, Chicago, IL: 8/18/08.

Steve Kent, Fermi National Accelerator Laboratory, Batavia, IL: 10/16/07; at Sloan Symposium, Chicago, IL: 8/18/08.

Jill Knapp, Princeton University, Princeton, NJ: 5/15/07, 1/15/08.

Rich Kron, University of Chicago; by phone, 6/27/07; at University of Chicago, Chicago, IL: 10/11/07.

Don Lamb, University of Chicago; by phone: 12/6/07.

French Leger, Fermi National Accelerator Laboratory; at Apache Point Observatory, Sunspot, NM: 7/5/07, 7/6/07; at Johns Hopkins University, Baltimore, MD: 4/16/08.

Chris Lintott, Oxford University; at Sloan Symposium, Chicago, IL: 8/16/08.

Craig Loomis, Princeton University, Princeton, NJ: 5/15/07.

Robert Lupton, Princeton University; at Institute for Advanced Study, Princeton, NJ, 5/17/07: 1/16/08.

Julie Lutz, University of Washington, Seattle, WA: 4/10/07.

Rachel Mandelbaum, Institute for Advanced Study; at Princeton University, Princeton, NJ: 5/14/07.

Ed Mannery, Walt Siegmund, and Russ Owen, University of Washington, Seattle, WA: 4/11/07.

Bruce Margon, University of California, Santa Cruz; by Skype: 8/22/08.

Kevin Marvel, American Astronomical Society; by phone: 5/23/07.

Tim McKay, University of Michigan; by phone: 10/25/07.

Matt Mountain, Space Telescope Science Institute, Baltimore, MD: 8/10/07.

Heidi Newberg, Rensselaer Polytechnic Institute; by phone: 12/21/07, 6/25/09.

Robert Nichol, University of Portsmouth; by e-mail: 5/21/08; by Skype: 5/21/08, 5/22/08; at collaboration meeting, Princeton, NJ: 7/27/09.

Jerry Ostriker, Princeton University, Princeton, NJ: 5/16/07, 8/14/07, 1/14/08.

Jim Peebles, Princeton University, Princeton, NJ: 8/15/07, 8/16/07; by phone: 7/24/09.

John Peoples, Fermi National Accelerator Laboratory, Batavia, IL, 10/15/07, 10/16/07.

Jeff Pier, National Science Foundation; at Sloan Symposium, Chicago, IL, 8/16/08.

Thomas Prince, California Institute of Technology, Pasadena, CA, 3/8/07.

Michael Richmond, Rochester Institute of Technology; by phone, 5/29/08.

Connie Rockosi, University of California, Santa Cruz; at Apache Point Observatory, Sunspot, NM: 7/9/07.

David Schlegel, Lawrence Berkeley National Laboratory; at Drexel University, Philadelphia, PA: 3/28/07; at Johns Hopkins University, Baltimore, MD: 4/18/08; at Sloan Symposium, Chicago, IL: 8/18/08; at collaboration meeting, Princeton, NJ: 7/28/09.

Don Schneider, Pennsylvania State University; by phone: 5/2/08.

Nigel Sharp, National Science Foundation; by phone: 3/6/08.

Sue Simkin, Michigan State University; by e-mail: 3/26/08, 3/27/08.

Robert Smith, University of Alberta; by phone: 4/10/08.

Michael Strauss, Princeton University, Princeton, NJ: 5/18/07; at Double-tree Inn, Baltimore, MD: 5/12/08, 5/13/08; by phone: 9/8/09.

Mark SubbaRao, Adler Planetarium/University of Chicago; by phone: 5/20/08.

Alex Szalay, Johns Hopkins University, Baltimore, MD: 2/21/07.

Ani Thakar, Johns Hopkins University, Baltimore, MD: 6/30/09.

Scot Tremaine, Institute for Advanced Study, Princeton, NJ: 5/17/07.

Michael Turner, University of Chicago, Chicago IL: 10/12/07; at Space Telescope Science Institute, Baltimore, MD: 5/7/08.

Tony Tyson, University of California, Davis; at Space Telescope Science Institute, Baltimore, MD: 5/6/08.

Alan Uomoto, University of California, Santa Cruz; at Johns Hopkins University, Baltimore, MD: 9/28/07.

Wayne van Citters, National Science Foundation, Arlington, VA: 2/27/08.

David Weinberg, Ohio State University; in Washington, D.C.: 12/24/07; at Sloan Symposium, Chicago, IL: 8/18/08; at collaboration meeting, Princeton, NJ: 7/27/09.

Simon White, Max Planck Institute; at Sloan Symposium, Chicago, IL: 8/16/08.

Curtis Wong, Microsoft Research, Redmond, WA: 4/10/07.

Brian Yanny, Fermi National Accelerator Laboratory; at Johns Hopkins University, Baltimore, MD: 6/15/09.

Don York, University of Chicago, Chicago, IL: 10/12/07.

Publications

Geoff Andersen. *The Telescope: Its History, Technology, and Future.* Princeton, NJ: Princeton University Press, 2007.

Eric J. Chaisson. *The Hubble Wars.* New York: HarperCollins, 1994.

Roger Davies, ed. *Instrumentation for Cosmology: A Summary of Contributions to the Workshop on Instrumentation for Cosmology, Held at NOAO (National Optical Astronomy Observatories), Tucson, February 18–20, 1987.* Tucson: NOAO, 1987.

Timothy Ferris. *Coming of Age in the Milky Way.* New York: William Morrow and Company, 1988.

James Gunn. "The 2.5 m Telescope of the Sloan Digital Sky Survey." *Astronomical Journal* 131 (April 2006): 2332.

Fred Hoyle. *Frontiers of Astronomy.* New York: Harper & Brothers, 1955.

Alan Lightman and Roberta Brawer. *Origins: The Lives and Worlds of Modern Cosmologists.* Cambridge, MA: Harvard University Press, 1990.

W. Patrick McCray. *Giant Telescopes: Astronomical Ambition and the Promise of Technology.* Cambridge, MA: Harvard University Press, 2004.

National Research Council. *The Decade of Discovery in Astronomy and Astrophysics.* Washington, D.C.: National Academy Press, 1991.

National Research Council. *Federal Funding of Astronomical Research.* Washington, D.C.: National Academy Press, 2000.

Richard Panek. *Seeing and Believing: How the Telescope Opened Our Eyes and Minds to the Heavens.* New York: Penguin Books, 1998.

Jim Peterson and Glen Mackie. "A Brief History of the Astrophysical Research Consortium and the Apache Point Observatory." *Journal of Astronomical History and Heritage* 9, no. 1 (2006): 109–18.

Richard Preston. *First Light: The Search for the Edge of the Universe*. New York: Random House, 1996.

Jean-Rene Roy and Matt Mountain. "The Evolving Sociology of Ground-Based Optical and Infrared Astronomy at the Start of the 21st Century." In A. Heck, ed. *Organizations and Strategies in Astronomy*, vol. 6. Springer, 2005.

Robert W. Smith. *The Space Telescope: A Study of NASA, Science, Technology, and Politics*. London: Cambridge University Press, 1993.

Michael Turner. "Quarks and the Cosmos." *Science* 315, no. 5808 (January 5, 2007): 59.

Joan Warnow-Blewett, Joel Genuth, and Spencer R. Weart. "Study of Multi-Institutional Collaborations: Final Report: Highlights and Project Documentations." http://www.aip.org/history/pubs/collabs/highlights.html.

Charles A, Whitney. *The Discovery of Our Galaxy*. New York: Alfred A. Knopf, 1971.

Sources for Specific Chapters

Prologue: The Instrument Fairy

Quotes from and opinions attributed to Jim Westphal come from: Interview with James A. Westphal. Oral History Project, California Institute of Technology Archives, Pasadena, California. Interviewer: Shirley K. Cohen, July 8–29, 1998: http://oralhisto ries.library.caltech.edu/107/01/OHO_Westphal _J.pdf (2002).

Jim Gunn's e-mail came from Don Schneider's electronic copy of it. The letter was addressed to Garth Illingworth, an astronomer at the University of California, Santa Cruz who was chairman of a panel for the 1990 decadal report of the National Academy of Sciences—the so-called Bahcall Report. Gunn was a member of

Illingworth's panel. He was responding to the panel's report, which recommended that the Space Telescope eventually be replaced and asserted that the astronomical community had been "screwed over" by the telescope's current problems. Gunn disagreed and accordingly wrote his letter. Gunn sent copies of his letter to John Bahcall and to Jim Westphal; Westphal excerpted the letter and sent it as an addendum to a status report on the space telescope to be sent to NASA.

Chapter 1: Stakes Worth Playing For

Quotes from and opinions attributed to Maarten Schmidt come from: Interview with Maarten Schmidt. Oral History Project, California Institute of Technology Archives, Pasadena, California. Interviewer: Shirley K. Cohen, April 11 and May 2 and 15, 1996: http://resolver.caltech.edu/CaltechOH:OH_Schmidt_M (1999).

The Herschel quote is from Whitney, *The Discovery of Our Galaxy*, p. 111; his drawing of the Milky Way is on p. 106. The description of Herschel's sweeps is in Ferris, *Coming of Age in the Milky Way*.

J. Richard Gott III, James E. Gunn, David N. Schramm, Beatrice M. Tinsley, "An Unbound Universe?," *The Astrophysical Journal* 194 (December 15, 1974): 543–53.

Information about CCDs came partly from James Janesick and Morley Blouke, "Sky on a Chip: The Fabulous CCD," *Sky & Telescope*, 74, no. 3 (September 1987): 238–42.

Chapter 3: Putting on the Play

The newspaper article announcing the survey was John Noble Wilford, "New Telescope to Enable Mapping of More Than a Million Galaxies," *The New York Times* (November 26, 1990): A17.

Chapter 5: Running Open Loop

Information on the 2dF survey comes partly from: http://www .mso.anu.edu.au/2dFGRS/.

Chapter 6: First Light

Some of the description of the AAS meeting came from James Glanz, "Astronomy: Giant Survey Wallpapers the Sky," *Science* 280, no. 5371 (June 1988): 1835.

Chapter 10: The Virtual Observatory

S. G. Djorgovski and R. Williams, "Virtual Observatory: From Concept to Implementation," *Astronomical Society of the Pacific Conference Series*, 345 (2005): 517. Available at astro-ph/0504006.

The quotes from and opinions attributed to Jim Gray are from a talk he gave to computer scientists at Johns Hopkins University on February 23, 2006.

Gray's idea of eScience is summarized in an article by Alexander Szalay and Jim Gray, "Science in an Exponential World," *Nature* 440, no. 7083 (March 23, 2006): 413–14.

The circumstances of Gray's early death are in an article in

Wired, July 23, 2007, available online at http://www.wired.com/techbiz/people/magazine/15-08/ff_jimgray?currentPage=all.

SkyServer is at http://cas.sdss.org. The Sloan's website is http://www.sdss.org/.

Galaxy Zoo is at http://zooniverse.org/home and http://www.galaxyzoo.org/. A nice article about Galaxy Zoo is: Devin Powell, "Amateur Hour," *Arts and Sciences Magazine* 5, no. 2 (spring 2008); http://krieger.jhu.edu/magazine/sp08/fl.html. The demographics of the zooites come from Jordan Raddick, 2010, personal communication.

WikiSky is at http://wikisky.org/. Google Sky is at http://www.google.com/sky/. Microsoft's World Wide Telescope is at http://www.worldwidetelescope.org/Home.aspx.

Sources for the virtual observatory are at http://www.us-vo.org/index.cfm, http://www.ivoa.net/, and http://www.astro.caltech.edu/~george/vo/.

The quote from Thomas Wright of Durham is from his "An Original Theory or New Hypothesis of the Universe," London, 1750; quoted in Dennis Richard Danielson, ed. *The Book of the Cosmos: Imagining the Universe from Heraclitus to Hawking*, Cambridge, MA: Perseus Publishing, p. 264. Background on Thomas Wright of Durham is in Whitney, *The Discovery of Our Galaxy*, pp. 77–86.

The idea that the archive's usefulness was limited only by the inventiveness of its users came from Zakamska, et al, "Challenges Facing Young Astrophysicists," State of Profession Paper for the Decadal Survey on astro-ph arXiv:0905.1986 (May 13, 2009).

Chapter 11: Everything's Different

The American Museum of Natural History's website on the Sloan is at http://www.amnh.org/sciencebulletins/astro/f/sdss .20051208/.

The information about the large surveys PanSTARRS and LSST came from their excellent websites: PanSTARRS, http://pan -starrs.ifa.hawaii.edu/public/, and LSST, http://www.lsst.org/lsst.

Chapter 12: Jim Again

The quote from Isaac Newton came from "Four Letters from Sir Isaac Newton to Doctor Richard Bentley Containing Some Arguments in Proof of a Deity" (between 1692 and 1756); quoted in Joseph Silk, "How Big Is the Universe," *Historical Development in Modern Cosmology*. Vicent J. Martínez, Virginia Trimble, and María Jesús, ed., ASP Conference Series 252 (2001): 117.

Some of the description of large-scale structure came from David M. Weinberg, "Astronomy: Mapping the Large-Scale Structure of the Universe," *Science* 309, no. 5734 (July 2005): 564–65.

Index

About the Author

ANN FINKBEINER is a freelance science writer who normally writes about cosmology and who runs the graduate program in science writing at Johns Hopkins University. Finkbeiner is the author of *After the Death of a Child* (Free Press, 1996) and *The Jasons*. She has written articles and book reviews for *Discover, Sky & Telescope, Astronomy, Science,* the *Wall Street Journal,* and the *New York Times,* as well as columns for *USA Today* and *Defense Technology International.*

Printed in the United States
By Bookmasters